INTERNATIONAL MINING FORUM 2009

18–21 FEBRUARY 2009
KRAKOW

Deep Mining Challenges

International Mining Forum 2009

Eugeniusz J. Sobczyk
Polish Academy of Sciences, Mineral and Energy Economy Research Institute, Cracow, Poland

Jerzy Kicki
AGH – University of Science and Technology, Department of Underground Mining, Cracow, Poland
Polish Academy of Sciences, Mineral and Energy Economy Research Institute, Cracow, Poland

Piotr Saługa
Polish Academy of Sciences, Mineral and Energy Economy Research Institute, Cracow, Poland

CRC Press
Taylor & Francis Group
Boca Raton London New York Leiden

CRC Press is an imprint of the
Taylor & Francis Group, an **informa** business
A BALKEMA BOOK

LONDON / LEIDEN / NEW YORK / PHILADELPHIA / SINGAPORE

CRC Press/Balkema is an imprint of the Taylor & Francis Group, an informa business

Typeset by Krzysztof Stachurski
Printed and bound in EU

Published by: CRC Press/Balkema
P.O. Box 447, 2300 AK Leiden, The Netherlands
e-mail: Pub.NL@taylorandfrancis.com
www.crcpress.com – www.taylorandfrancis.co.uk – www.balkema.nl

ISBN: 978-0-415-80428-8

International Mining Forum 2009, Sobczyk, Kicki & Saługa (eds) © 2009 Taylor & Francis Group, London, UK, ISBN 978-0-415-80428-8

Table of Contents

International Mining Forum 2009, Sobczyk, Kicki & Saługa (eds) © 2009 Taylor & Francis Group, London, UK, ISBN 978-0-415-80428-8

Preface

Mining of 21st century will face the problems of mineral resources depletion and searching for possibilities of developing deposits that today cannot be mined technically and economically. One of the most actual problems is then extraction at greater depths. It will create challenges for mining. Fortunately, physical limits to mining are tested again and again until all the insurmountable obstacles are finally overcome. Nowadays, through technological advances, shafts reaching over 3,000 metres are possible, so the question is – where are these physical limits?

Accessing reserves at depth must face with breaking down various barriers. The first ones are ground control issues resulting from high stress hazard; they are mainly permanent infrastructure stability problems, better understanding of rock mass behaviour, new mining methods that could help avoid or minimize rock burst occurrence, better engineered backfill to withstand the pressure from the convergence of mine openings, etc. High pressure issues together with the conditions of the rock mass, including weaknesses, watercourses, faults and joints, and ambient rock stresses, result in "rockbursts" (in hard rock mining technology) or "bumps" or "outbursts" (in coal mining terminology). The second barrier is the problem with underground environment; it focuses on dealing with heat from strata and equipment, humidity, and noxious gases in the mine atmosphere causing expensive air-conditioning and ventilation capacity. It can be assumed that energy cost will increase steadily for the foreseeable future. In some mines ventilation and air conditioning represents today 50% of the energy required (one of the reasons is that ventilation costs increase exponentially with demand). One is looking for new engineering tools to enable low emission and low heat generation. Mentioned tasks create the next barrier – the ability and willingness of people to work under so inconvenient conditions underground. This forces searching for new mining methods that are suitable for extreme environments. The next barrier is logistic – great depths involve transportation, rock and material handling, and communication problems result in high cost, loss of effective working time and difficult inventory control.

International Mining Forum 2009 works confirm that safe and economical development of new deep orebodies in existing mines and possible opening of new mines is possible. The book presents some other texts presented in the Forum concerning mining technology and economy that complete this edition.

Eugeniusz J. Sobczyk, Jerzy Kicki & Piotr Saluga (eds)

International Mining Forum 2009, Sobczyk, Kicki & Saługa (eds) © *2009 Taylor & Francis Group, London, UK, ISBN 978-0-415-80428-8*

Organization

Organizing Committee:
Jerzy Kicki (Chairman)
Eugeniusz J. Sobczyk (Secretary)
Artur Dyczko
Jacek Jarosz
Piotr Saługa (Secretary General)
Krzysztof Stachurski
Agnieszka Stopkowicz
Michał Kopacz

Advisory Group:
Volodymyr Bondarenko, professor (National Mining University of Dnipropetrovsk, Ukraine)
Wojciech Bradecki (Poland)
Jan Butra, professor (Wroclaw University of Technology, KGHM CUPRUM Ltd., Poland)
Alfonso Carvajal, Ph.D. (Universidad de La Serena, Chile)
Piotr Czaja, professor (AGH – University of Science and Technology, Poland)
Jozef Dubinski, professor (Central Mining Institute, Poland)
Jaroslav Dvoracek, professor (Technical University VSB, Czech Republic)
Pawel Krzystolik, professor ("Barbara" Experimental Mine, Poland)
Garry Litwinsky, professor (Donbass State Technical University of Alchevsk, Ukraine)
Eugeniusz Mokrzycki, professor (MEERI of Polish Academy of Sciences, Poland)
Roman Ney, professor (MEERI of Polish Academy of Sciences, Poland)
Jacek Paraszczak, professor (Laval University, Quebec City, Canada)
Janusz Roszkowski, professor (AGH – University of Science and Technology, Poland)
Stanislaw Speczik, professor (Polish Geological Institute, Poland)
Anton Sroka, professor (Technical University Bergakademie Freiberg, FR Germany)
Mladen Stjepanowic, professor (University of Belgrade, Republic of Serbia)
Antoni Tajdus, professor (AGH – University of Science and Technology, Poland)
Kot F. v. Unrug, professor (University of Kentucky, USA)
Leszek Wojno, Ph.D. (Australia)

Deep Mining Challenges

International Mining Forum 2009, Sobczyk, Kicki & Saługa (eds) © 2009 Taylor & Francis Group, London, UK, ISBN 978-0-415-80428-8

Structure and Technological Elements of Information System of Mine Based on the Control of a Condition of Rock Mass

Volodymyr I. Bondarenko, Volodymyr I. Buzylo, Vadym V. Fomichov
National Mining University, Dnipropetrovs'k, Ukraine

ABSTRACT: The technique of formation of information system of mine based on the account of a geomechanical condition of rock mass of the coal-mining enterprise containing underground construction is presented.

KEYWORDS: Information system, a rock mass, mine, production management, geomechanics

The majority of mines of Ukraine function in difficult rock-geological conditions. Therefore for these mines it is impossible to use the usual approach to working out and introduction of an information control system by manufacture. Such information system should possess a number of features which will allow to operate effectively coal mining process in difficult and emergencies. Such situational management is not possible without a current condition of rock mass in which limits underground constructions of mine settle down.

The analysis of information systems already applied in branch and their analogues offered in the software market, has allowed to reveal following basic lacks: information systems do not cover completely all technological chains of mine; are not capable to function in the situations which description is not provided in the basic technological process; demand constant intervention of the operator; possess great values of time of reaction and poor quality of integration at level of the centralised management of association of the mining and processing enterprises.

The basic part of these problems can be solved in the course of designing of information system of mine having entered in it possibility of the account of features of behaviour of rock mass. At extraction of a mineral the rock mass condition can cardinally change for a short time interval. Therefore definition of a condition of rock mass in real time to become a major principle of designing of information system of mine. Such control system demands special logic structure capable to realise a major principle of designing of system in full and not at the expense of reduction of functionality of tasks of second importance of management.

It is possible to present the general logic structure of information system of mine in the form of three-level hierarchy (Figure 1): I – level of logic blocks; II – level of logic contours; III – level of initial problems. Each logic block possesses own operational kernel independent of other blocks. Logic contours unite initial problems by a principle of technological interrelations. The contour completely disappears in the logic block also can be realised as an independent computing stream within one operational environment. Level of initial problems contains a set of independent objects providing functional connection between an information control system and elements of real technological process. Each initial problem represents an independent program code realising one procedure of management.

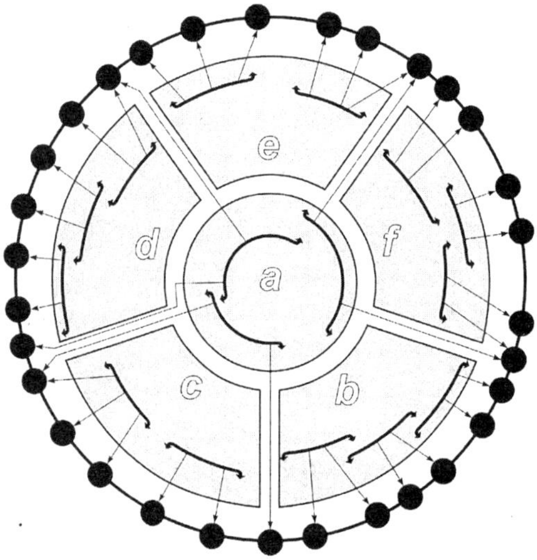

Figure 1. The general logic structure of information system of mine

At level of logic blocks the structure of an information control system of mine consists of following elements (Figure 1): *a* – the management mainframe; *b* – the block of management of databases; *c* – the emergency control block; *d* – the reception and data transmission block; *e* – the block of connection of public networks; *f* – the block of interfaces of management. Each block can contain from two and more logic contours each of which in turn provides access to initial problems. All logic contours uniting some initial problems exclusively operate them. If there is a necessity to get access to initial problem of other contour it is possible to execute only at intermediary of the mainframe of management. Its logic contours have possibility of direct access to any elements of a control system of mine.

On an external circle of the scheme initial problems are located. Their number defines the general complexity of a control system. The number of initial problems a part of a concrete logic contour defines accuracy of the description and level of controllability of certain technological process. Each initial problem is unique within one control system of mine. Physically initial problem can serve some devices of one type. Such approach is realised by creation of a separate copy of an initial problem for each device connected to information system.

Within the limits of given article we will most in detail consider the management mainframe, the block of an emergency control and the reception and data transmission block. We will separately consider questions of interaction of these elements of information system of mine.

The management mainframe is realised in the form of the multiline allocated computer complex which structure is presented in Drawing 2. The main element of the mainframe of management is system the kernel. Its functions include providing of access of a program code to computing powers of system, distribution of time of processors and data transmission maintenance between logic blocks of system and its separate interfaces.

Contours of the mainframe of management function in an exclusive mode of access to resources of a system kernel. In Drawing 2 it is visible how these contours can co-operate with each other. There are two possible modes one data. It is the standard report of an exchange at intermediary of a system kernel and direct data exchange.

The standard report of data exchange is used by a system kernel as the uniform logic interface at service of inquiries of any elements of a control system by mine. Reports of direct data exchange can have various structures for different pairs of elements of information system. These reports are used between contours of system in need of high-speed data exchange or at high complexity of structure of transferred data.

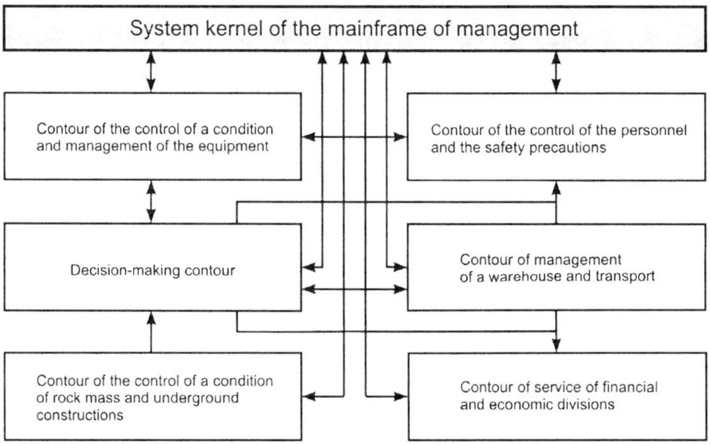

Figure 2. Structure of the mainframe of management

The decision-making contour is the most important of elements of a control system of mine after a system kernel. This contour possesses two unique function. The contour is capable to accept data and to transfer commands to execution to initial problems bypassing logic contours to which they belong. Possesses the privilege of blocking of commands of interfaces of management arriving from the block. It is necessary for decrease in industrial risk connected with the human factor that is especially actual for manufactures with the raised level of breakdown susceptibility.

The contour of decision-making makes the analysis of data transferred of other logic contours and blocks of a control system of mine. If the data set received by a control system cannot be processed in the standard image or there is a situation demanding a choice of a line of conduct the decision is accepted by a decision-making contour. This contour the unique part of a control system which can dynamically change the program code under control of a contour of self-training of a control system of the block of an emergency control.

The contour of the control of a condition and management of the equipment carries out remote control by all kinds of the equipment of a mineral used at extraction. Initial problems of a contour exercise administration concrete devices. These devices within a contour unite in technological chains for which the separate logic problem of management dares. This contour is the basic consumer of computing powers of a hardware complex of a control system of mine and completely functions in a mode of real time.

The contour of the control of a condition of rock mass and underground constructions carries out monitoring of a condition of underground mine working, them supports, atmospheres of mine and deformations and pressure surround rock mass. As the given module the analysis of changes of controllable indicators and delivery of recommendations about the organisation of operation of underground constructions of mine goes.

The contour of management of a warehouse and transport carries out the control over moving and storage of cargoes and the extracted mineral. Management of a warehouse carries out a contour directly. Management of transport is divided into two parts: management of cargo streams (internal and external logistics) and management of the transport equipment. Management of the transport equipment carries out the given contour through intermediary of a contour of the control of a condition and management of the equipment. The decision of problems of internal and external logistics is realised in the form of two separate sets of initial problems which joining carried out at contour level.

The contour of service of financial and economic divisions is used for the organisation of the general management by mine divisions. This contour functionally completely is independent of other elements of information system of mine. Data exchange between it a contour and other elements of information system of mine occurs through databases of a control system or at intermediary of the block of connection of public networks.

The contour of the control of the personnel and the safety precautions in territory of an industrial complex of mine solves two primary goals: supervision over personnel moving; the organisation of performance of requirements of safety precautions and salvage operations. The contour allows visually and by means of system of gauges to define a position of any miner and by means of various devices of the alarm system to operate its moving within an industrial complex of mine. By means of initial problems of a contour evacuation of miners in case of occurrence of emergencies menacing to their life will be organised.

Separate and very important problem is management of mine in emergencies. At occurrence of such situations it is necessary to provide possibility of operative rescue of miners and to minimise possible material losses of manufacture. Therefore in structure of a control system the mine includes the emergency control block (Figure 3). This block works as different image in two modes set for a control system: standard and emergency.

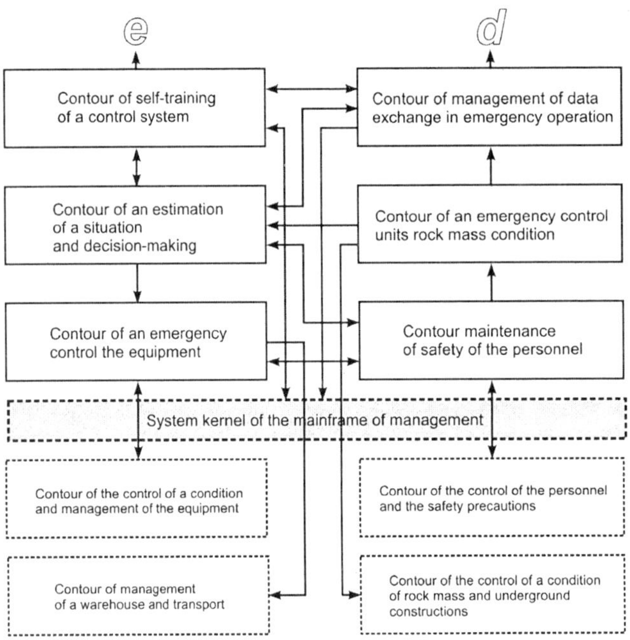

Figure 3. Structure of the block of an emergency control and its interaction with the management mainframe

The emergency control block in a standard operating mode of a control system of mine carries out auxiliary functions: operates work of reserve system; solves problems on self-training of system by the analysis of data arriving from others logic block; and in a mode of real time estimates a condition of technological processes of mine.

At occurrence of a supernumerary situation the emergency control block estimates degree of danger of the arisen problem and carries it to one of three categories: dot failure; local failure; failure of a technological chain; general average. The category choice defines degree of intervention of the block of an emergency control in a standard operating mode of a control system of mine. The emergency control block can transmit only one command or completely disconnect and replace a part of the logic block of system.

The contour of an estimation of a situation and decision-making in a standard mode performs function of duplication of work of the basic logic blocks of a control system by mine. In case of emergency occurrence this contour from a background mode passes in a mode of the foreground and provides transfer of a part of functions of logic blocks of system to emergency contours.

The contour of self-training of a control system in a standard operating mode of a control system of mine makes the analysis of reactions of a control system of mine on various industrial situations. As the information for the analysis the given estimations of a situation processed by a contour and decision-making serve. In case of the negative analysis of reaction of a control system the self-training contour forms a new program code and introduces it at intermediary of a system kernel in a contour of decision-making of the mainframe of management.

Contour of an emergency control units of a condition of rock mass, a contour of management as data exchange in emergency operation, a contour of an emergency control the equipment and a contour of maintenance of safety of the personnel in a standard mode do not work and can be used a contour of self-training of a control system for modelling of various industrial situations.

At transition in emergency operation of work of a control system by mine the dominating role in system is received by the emergency control block (Figure 4). Transition in emergency operation of a control system by mine initiates the emergency control block. In this mode a part of contours of this block change the scheme of functioning.

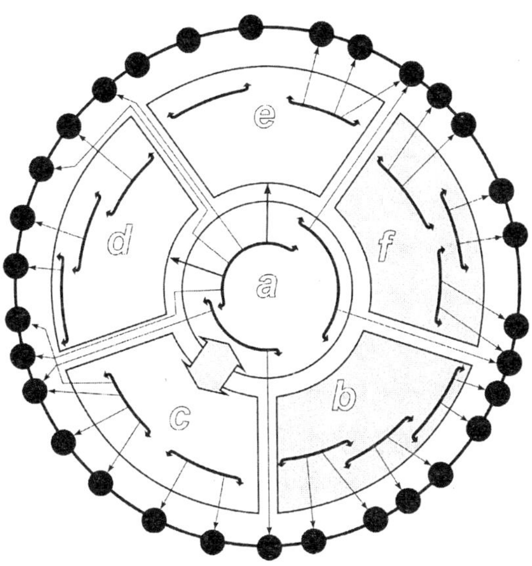

Figure 4. The general logic structure of information system of mine at emergency occurrence

The estimation and decision-making contour completely replaces with itself a contour of decision-making of the mainframe of management and passes in a verification mode of all inquiries transferred in a system kernel of the mainframe of management.

In emergency operation to become the basic function of a contour of self-training of a control system maintenance of reserve restoration of a condition of computing processes in case of hardware or logic failure of a control system. Besides, this contour at intermediary of a system kernel of the mainframe of management changes conditions of functioning of the block of public networks.

The contour of an emergency control the equipment replaces with itself two contours of the mainframe of management (a contour of the control of a condition and management of the equipment; a contour of management of a warehouse and transport). The contour of emergency control units of a condition of rock mass changes an operating mode of a contour of the control of a condition of rock mass and underground constructions of the mainframe of management. In a similar way the contour of maintenance of safety the person interferes with work of a contour of the control of the personnel and the safety precautions and disconnecting a part of its functions uses data of this contour for planning of actions for rescue of people and material assets.

As a whole emergency occurrence leads to change of an operating mode of four logic blocks of a control system by mine (Figure 4), that makes more than 80 % of its structure.

Efficient control technological elements of production of mine probably only at maintenance of high quality and speed of an information exchange between the basic logic modules of information system. The big number of various hardware interfaces, the big distances of data transmission and high level of hindrances are serious problems of maintenance of qualitative and fast data transmission in an underground complex of mine. The reception and data transmission block is engaged in the decision of these problems.

This block represents set of contours providing data exchange procedures between devices with various hardware interfaces and program modules of system. Each contour serves certain type of the report of data exchange on the internal computer network of the enterprise (Figure 5).

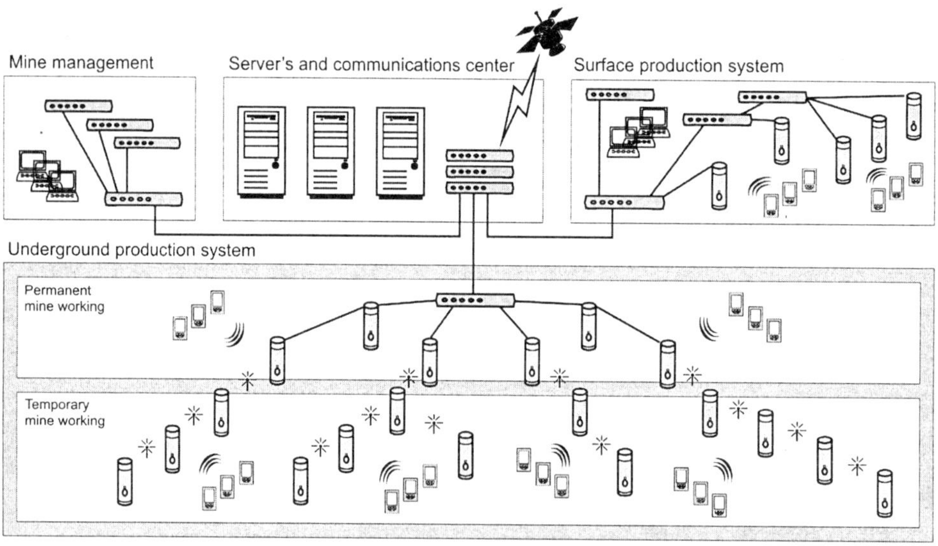

Figure 5. The scheme of the organisation of a local network of mine

Network devices and hardware interfaces, can exchange data through a local network of the enterprise only at intermediary of copies of initial problems belonging to one contour of the block of reception and data transmission. There is no sense to describe separately each contour of this block as all of them carry out the same functions. The number of contours depends on number of types of network interfaces used in a concrete configuration of a control system of mine.

Into problems of the block of reception and data transmission enters: maintenance of reception of data from program modules of a control system of mine; the analysis and transformation of these data; transfer of the transformed data to the concrete hardware interface; maintenance of return delivery of data from the hardware interface to the program module; the control over a condition and management of the network equipment; check and an estimation of a condition of the environment of transfer of a digital signal.

All initial problems of the block of reception and data transmission are carried out distantly, on the equipment located within a land and underground industrial complex of mine. Thus, the pre-processing of signals of the hardware interface occurs without their transfer to a local network of mine. As a result all information interchange in a local network of mine is carried out by means of the special simplified network report that allows to raise reliability and speed of data transmission. Such approach is possible only at use of special network devices of reception and data transmission.

The local network of mine provides data exchange between groups of servers both stationary and mobile digital terminals. As Wednesday of transfer copper twisted pair, an optical fibre and radio-waves is used. Stationary terminals settle down in a land industrial complex, in management of mine and capital mine working of an underground industrial complex. Such terminals join the computer network by means of standard switchboards using as Wednesday of transfer twisted pair or an optical fiber. In temporary mine working for some reasons to organise the computer network in a similar way it is not meaningful.

For the organisation of a local network of mine in temporary mine working use the special compact modular device. There is an acquisition possibility its different sets of interfaces of reception and data transmission. This device can be used as a standard repeater of a radio signal, as the wireless network interface of the industrial equipment and as the bridge between two environments of data transmission (Figure 5). Besides, as the given device providing of access to a local network of mine of digital mobile terminals in a land and underground industrial complex of mine enters.

Without application of the similar device work of the block of reception and data transmission is not possible. The part of functions of this block is realised directly within local architecture of the given device. Such functions concern: the analysis, transformation and compression of data; check of availability of the environment of data transmission and its productivity, the control of topology of the computer network in temporary mine working.

Besides, these devices in certain configurations are used by a contour of the control of a condition of rock mass and underground constructions. Supplied with various gauges and the alarm equipment these devices allow to provide with data all contours of the mainframe of management.

The described logic structure and separate technological components of information system of mine allow taking into account change of a condition of rock mass in real time effectively to operate coal mining process even in difficult mountain-geological conditions and at occurrence of emergencies. The mine control system constructed on the offered principles will provide high a production management degree of quality will lower personal risks of miners and will allow to minimise influence of the human factor at management of technological process.

International Mining Forum 2009, Sobczyk, Kicki & Saługa (eds) © 2009 Taylor & Francis Group, London, UK, ISBN 978-0-415-80428-8

Methane Hazard Predictions in Underground Coal Mining

Marek Borowski, Nikodem Szlązak, Dariusz Obracaj
AGH – University of Science & Technology, Krakow, Poland

ABSTRACT: Methane emissions present a major hazard during underground coal mining. The knowledge of the emission levels is required to plan the mining operations, to select the mining method and the degasification strategy. Methane emission levels can be determined from reliable forecasts. Prediction methods involve a number of models which vary in the degree of precision. The study outlines the empirical methods, followed by a prognosis of methane emissions based on the geomechanical model. Simulation methods and computer programs are presented that are used to evaluate the methane hazard. Prediction methods used in Poland are reviewed in further sections. One has to bear in mind, however, that these methods may not be effective in deep mines. Finally, the forecasting method is presented that is based on neural networks.

INTRODUCTION

Coal mining operations at increased depths are continued in difficult geological and mining conditions further threatened by natural hazards, one of the major hazards being the methane emissions associated with the presence of methane in the rock strata and its release during the mining operations.

Methane predictions aim to evaluate the risk of methane emission (absolute or relative methane bearing capacity) as accurately as possible, for a newly constructed mine, a mining level being developed or for a new ventilation region; in particular the methane emission hazard has to be evaluated for each mine working or a longwall region. One has to bear in mind, however, that all forecasts are only approximations of the conditions that might arise when mining operations are commenced.

Methane prediction methods include empirical, analytical and statistical models and their combinations. Forecasts might cover various time periods: several days, several weeks or stretch just to the nearest future. They are related to the daily output levels or to the daily face progress. Some predication models involve methane emissions from longwalls, others include also methane coming from the galleries in the neighbourhood of the face region.

Most methods enable the forecasting of methane emissions in steady-state conditions, when methane inflow is constant and the dynamics of change can be neglected. The majority of prediction methods do not handle unexpected methane release to the mine workings and no method is available yet that should adequately handle this phenomenon. Luckily, newest computer programs afford us the means to treat those processes jointly.

Forecasting of methane emissions in mines is based on:
- empirical models,
- simulation models,
- short-term forecasting (pseudo-real) models,

- computer assisted methane control systems (operating in real time and supporting the monitoring systems in mines).

All these models allow the methane emissions to be forecasted, moreover in some cases they help to identify the potential control measures or to design effective methane drainage systems.

EMPIRICAL MODELS

Empirical models of methane emissions have been developed in several countries: Belgium –INIEX (Institute National des Industries Extractives), France – CERCHAR (Centre d'Etudes et de Recherches des Charbonnages de France), Germany – STVB and WBK (Steinkohlenbergbauverein and Westfalische Berggeverkschaftskasse), Poland – in an test mine Barbara, Great Britain – MRDE (Mining Research and Development Establishment) and in former Soviet Union – the Skoczynski Institute. These are mathematically simple methods, requiring several input parameters and are usually related to conditions specific to the given coalfield. The methods are able to handle methane emissions from the following sources:

- coalbed methane,
- overlying and underlying strata,
- the exposed longwall face surface,
- coal being transported on conveyors.

Most forecasts assume that methane is released from strata disturbed by mining operations up to 200 m above and 70 m below the coal seam being mined. Prediction methods assume that coalbed methane is the main source of methane in the mines, another source of emission are waste rocks, particularly sandstone.

These prediction methods are based on similar assumptions. First of all, the methane emission zone is determined in the longwall face region, referred to as the desorption zone, and that parameter is used to calculate the emission levels at some distance from the coal seam being mined. Typically, emission levels depend on the primary gas content in rocks (gas bearing capacity). The emission levels from particular coal seams and gas inflows are associated with the methane contents in the coal seam and the output from the given longwall (Noack, Opahle 1992).

A slightly different method, known as MRDE, is widely employed in Great Britain. The method, taking into account the effects of time in the mining operations, uses mathematical descriptions of gas emissions to the mine workings (Airey 1971). The emission levels of the mine gas from the coal seam being mined as well as the roof and floor strata are considered in the variable state of stress of the rock strata. Methane emissions to the longwall face region is given as a function of gas bearing capacity of the coal seam being mined, the number and thickness of coal seams in the region disturbed by mining operations, the distance between the disturbed seams and the coal seam being mined, the time of longwall operation from its commencement and the face advance rate.

Methane prediction methods have been tested in Australia (Jensen et al. 1992) and received a great deal of criticism for too many simplifications and too small number of variables used in most complicated forecasting. In their publication the authors listed 36 variables that influence the level of methane emissions to mine workings. Theoretical backgrounds were formulated correctly; though obtaining such a great number of data for predictions presents a major difficulty. When the data are insufficient, however, the forecasts will not be reliable.

GEOMECHANICAL MODEL OF METHANE HAZARD PREDICTION

The prediction method underlying the computer programs "Floorgas and Roofgas Geomechanical and Gas Release Model" described by Lunarzewski (Lunarzewski et al. 1995b) is used for forecasting methane emissions to mine workings, taking into account the involved geomechanical and geological processes. The method uses computer programs to design the gas recovery methods to be employed both on the surface and underground. They are also used to determine the amounts of methane released and emitted to mine workings and to evaluate the required ventilation conditions. Both the amount (proportions) of gas released during the mining operations and the actual shape of the gas level curve in the gobs are governed by the mathematical formula derived for highly efficient longwall faces (Lunarzewski 1992a; Lunarzewski 1992b).

The method makes use of a number of geomechanical parameters: depth and thickness of the coal seam being mined, mechanical properties of floor and roof strata, compressive strength of rocks or general classification of rocks: strong or weak, speed of sound in the rock strata. Stress is expressed as the horizontal pressure and horizontal to vertical stress ratio. The angle is considered between horizontal stresses and the face axes directions. Lithological data are required to define the properties of roof and floor strata and those of the coal seam being mined as well as for roof strata above 200 m and floor strata below 100 m below the mined coal seam. The relationship is utilised between rock properties and the speed of sound in the rock strata. Finally, some mining data are required: geometry of the coal seam being mined, longwall face advance or output levels, intensity of gas release, gas pressure or other emission parameters.

Computer programs use the relationship between strata de-stressing and the gas emission phenomenon. The program "Floorgas" combines the detailed rock mechanics analysis with gas capacity studies to compute the stress zones and gas release zones in the floor of the coal seam being mined. The simulation program "Floorgas" calculates the existing state of stress in the rock strata by the FEM method, performs the superposition of horizontal tectonic stresses, finds the shearing zones where gas may be released and generates the de-stressing angles, de-stressing zones and their boundaries in the rock strata. The program generates a vertical function of load distribution along the face range over the virgin coal seams and over the selected caved-in section, as shown in Figure 1. On that basis the vertical loading is computed for each cubic metre element of the modelled strata.

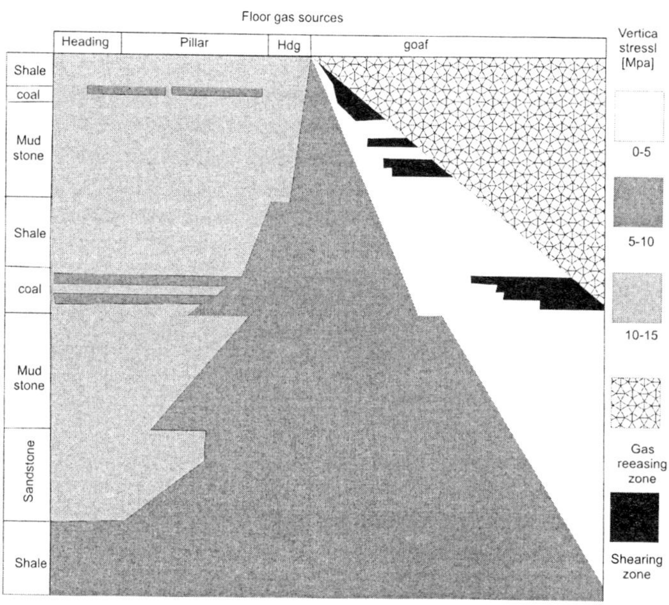

Figure 1. De-stressing zones in the floor of the coal seam being mined (Lunarzewski et al. 1995b)

The simulation program Roofgas utilises the rules of sequential separation of coal seams to derive de-stressing data, the amounts of released gas and local zones of discontinuous deformations in the roof. The program generates the roof fracture line as the boundary between continuously and discontinuously structured rocks. Besides, it generates de-stress angles and five positions in the gas release zone from the roof strata, defining the shape of these five zones for various degasification intensities (Fig. 2), in accordance with the empirical model (Lunarzewski 1992a).

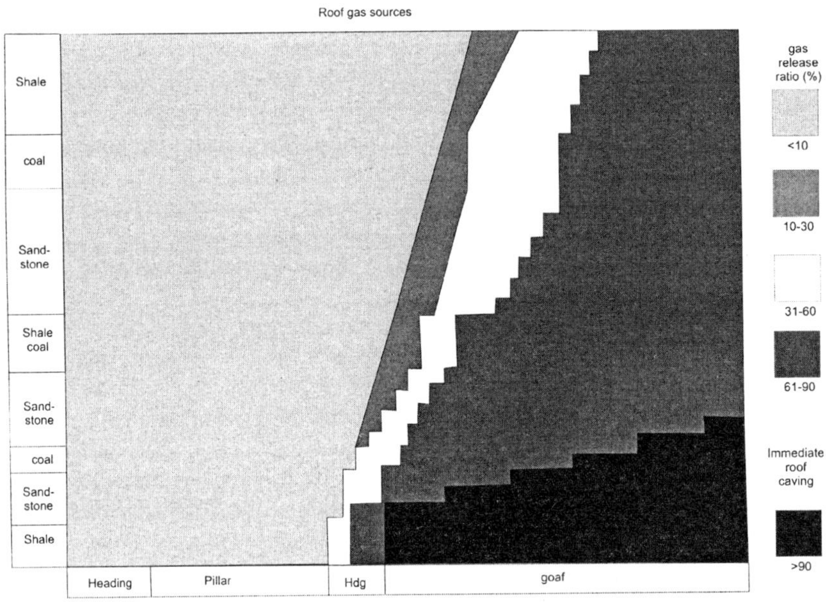

Figure 2. De-stressing zones in the roof of the mined strata (Lunarzewski 1995b)

Computer programs enable us to determine gas emissions during the longwall operations and to find the relationship between gas release, longwall face length and coal output levels. Optimisation of length, location and inclination angles of the goaf boreholes as well as the length, location and inclination angles of boreholes in the coal seam being mined. Computer programs generate vertical cross-sections through rock strata for selected distances ahead and behind the operational longwall. Floorgas shows de-stressing zones up to 100 m below the mined coal seam. The program Roofgas allows the de-stressing zones to be located 200 above the coal seam being mined.

Computer-assisted gas bearing capacity forecasts were tested in the conditions of high productivity operational mines in Australia and elsewhere, during the period 1992–1998. When the input data were accurate, the relative error involved in methane emission forecasts would approach 20%, which was confirmed in Australian mines. In terms of their form and contents, the computation data seem more adequate to evaluations of methane drainage schemes than to methane emission forecasts in mines.

Simulation models

Simulations of methane flow are widely employed to assist in coal mining and in methane recovery for commercial use. Simulation models yield the gas flows in the coal seam and hence can be utilised in design and evaluation of degasification schemes. These models largely emulate the real life conditions, though they display a high level of complexity. One of the most advanced model (Durucan et al. 1993) takes into account the dual porosity (of coal and rocks), dual permeability (gas and water) and the influence of stress on porosity and permeability.

Saghafi (Saghafi 1989; Saghafi et al. 1987) investigated unbalanced desorption processes in relation to coalbed methane emissions. The model is used for 2D calculations by the finite difference method. The relationships between rock stress, pore pressure, de-watering and coal seam permeability are duly considered.

Patton (Patton et al. 1994) developed a model whereby a coal seam is treated as reservoir and gas flow is simulated taking into account the drainage holes in the coal seam and gas release in the from the exposed longwall face. Air flow in the ventilation network is considered, too.

All these simulation models are used by their authors for research purposes and are not available commercially. There are some commercially available programs on the market, though their applicability in the mines is vastly limited as they fail to handle methane emissions from different sources, for instance coalbed methane and methane coming from gobs or abandoned workings.

Coalgas, developed by S.A. Holditch and Associates (SAH) is the only commercially available simulation program which permits the calculations of methane emissions from gobs. Accordingly, a coal seam is divided into discrete blocks, all of which form a reservoir. Each block is characterised by the following properties: permeability, porosity, water content, the amount of absorbed gas, also incorporating the available mining and geological data. Blocks represent coal present in a coal seam that might be removed in the course of mining operations within the predetermined period of time, without the need to re-start the simulation procedure. Each face displacement is taken into account. Gas released from each removed coal block is computed to determine the amount of incoming gas, which flows to the longwall region during the mining operations. Such simulation procedure enables us to determine the ventilation conditions in the longwall face zone and to select the appropriate degasification scheme.

The CFD approach is widely employed in simulations of ventilation processes, including gas emissions and gas flows in gobs. Calculations were performed in the Nottingham University (Moll et al. 1994). However, CFD methods fail to provide technical solutions required computing methane emissions; they only help in evaluations of the problem.

SHORT-TERM MODELS OF METHANE HAZARD PREDICTION

Short-terms models can be hardly classified as methane emission forecasts; they are rather observations of mining and gas release processes. This main purpose of such observations is to identify potential methane hazard during mining operations, taking into account the face advance rate or the face output levels, by comparing the prognosticated and real values. Observations are now possible thanks to close monitoring of the mining data.

Experiments conducted in MRDE (Great Britain) involved the correlations between the average daily methane emissions and daily output from a longwall face. However, no sufficiently accurate correlation was found. It was only after the coal production history was considered some time back, that the adequate results were obtained (CEC 1989). Besides, the algorithm allowed for prognosticating methane emissions in steady-state conditions. However, in the conditions of more intensive methane inflow, the results were far from satisfactory.

Statistical approach to predictions of methane emissions on a weekly basis was applied in the coal basin in Lotharingia, in France (Tauziede, Pokryszka 1993). In this case the correlation was found between the amount of released methane and the weekly face advance. However, when the bi-weekly periods were concerned, the results proved less accurate and the method was unsuccessful when the beginning of the face range was considered (up to about 200 m).

The influence of longwall face production levels and methane emissions was rigorously studied in an experimental mine Barbara (Cybulski, Myszor). The experimental program was extensive, covering 21 faces in six collieries. It involved measurements of methane concentrations taken with automatic recording manometers, in selected faces measurements were also taken of airflow velocity using automatic recording anemometers. The test procedure in each longwall region lasted for about 7 months. Measurement data were then subject to statistical treatment. It appears that as the output level increases, methane emission initially grows fast, and then slows down and the relationship between methane emission and the output might be well approximated by a power function. Even though thus derived formula seemed universal enough, it was still applicable only in such conditions as prevail in the Upper Silesia Coalfields.

Potential applications of time series to the modelling of variations of methane concentrations in coal mines were explored, too (Barker-Read, Radchenko 1989). The correlation was studied between the amounts of methane captured in the drainage systems and variations in methane concentrations in mine workings. The analysis of time series reveals that this method cannot well handle very slow and very fast fluctuations of gas concentrations. However, the application of a multivariance model proved most promising when the parameters required for predictions are available on a continuous basis. This system proved adequate in control of methane concentration parameters in the degassification systems.

Dixon and Longson (Dixon, Longson 1993) applied the computational procedure involving time series based on the multi-variance model to the forecasting of hourly fluctuations of methane concentration in a ventilation gallery, having taken account of the output levels. The model takes into account all sources of methane: coalbed methane, the exposed longwall face, coal transports, and methane permeating from the strata below and above in the longer periods of time. However, account is not taken of the effects of airflow rate variations, which is absolutely essential in forecasting. That is why this model cannot be regarded as reliable.

Badura (Badura 2001) attempted to develop a short-term forecast of methane emissions in the longwall face region based on the time series theory and using the available data from the mine. The analysis of methane emissions in the function of time based on autocorrelation involving additive and multiplicative methods reveals that the time series of methane emission from the area displays periodic/seasonal fluctuations, the season length being 7 days. Among four trend functions used in the calculations, no function was found that would yield satisfactory results throughout the whole cycle of longwall operation or that would give best results in the considered period of time.

Methods discussed so far are not helpful in finding the amounts of methane released to mine workings. Yet the state-of-the-art manufacturing control and monitoring systems provide a vast body of data which might be further used to evaluate methane hazard.

SYSTEMS SUPPORTING VENTILATION PARAMETERS CONTROL

During the coal mining operations methane concentration tends to increase and methane emission predictions might be based on mathematical modelling. When mining operations are continued, methane hazard might occur when ventilation conditions are changed (for instance when dams are open), when methane is pushed from gobs after cave-ins, when barometric pressure changes quickly or when methane flows from behind the dams. Most forecasts fail to take account of these conditions, moreover, we do not know the exact time and place of their occurrence and that is why their impacts cannot be precisely assessed.

Computer programs, which can be classified as expert systems, do not emulate the full repository of human knowledge and no account is taken of random factors. These computer programs will manipulate the knowledge that is available to humans rather than observation data. They are well suited to evaluating processes that are either incorrectly defined or too complex to permit a full mathematical analysis. Table 1 summarises computer programs that are in widespread use in various countries.

The program for methane drainage control EFEL was developed in Great Britain (CEC 1988) and supplemented with forecasts of methane release into mine workings. It was designed to support the predictions of likelihood of methane release to mine workings. The program shows the parameters having relevance to potential mining hazard. Incorporation of geological, mining and ventilation conditions gives us a better insight into gas emission processes and their causes.

Table 1. Computer programs used in different countries

Name	Country	Developed by	Approximate date	Applications
METHPRO	USA	USBM 1987	1986/7	Methane hazard control in mines
UKMVM	USA	University of Kentucky	1989	Control of ventilation parameters, data monitoring and analysis
UFEL	UK	British Coal CEC, 1988	1986	Evaluation and minimisation of gas emissions in British mines
HELPDRAIN	UK	British Coal CEC, 1989	1990	Methane hazard prediction, methods and control
BURST	China	Yansheng D. et al., 1990	1990	Gas emission forecast, gas outburst hazard, control methods
PROGMET	Poland	IMG PAN Kraków	2000	Methane concentration predictions in ventilation networks

A similar approach was adopted by USBM that created the METHPRO program used in methane hazard monitoring and control. Major advancements in measurements and recording of physical parameters: methane concentration, airflow velocity, supporting the monitoring systems in mines enabled the development of computer programs operating in real time and interacting with monitoring systems in mines. An example here is the program PROGMERT operating in real time and integrated with the monitoring system in the mine (Dziurzyński et al. 2002). Monitoring sensors measure methane concentrations in several points in the mines on a continuous basis. Application of the program facilitates the recording of monitoring data (methane emissions, velocity, pressure) to predict methane emissions in the context of ventilation conditions.

Computer programs are most helpful in searching for the optimal methane control solution. To solve this complex problem, a methane hazard prediction system is required to incorporate other natural hazards that might occur during mining operations, i.e. rock bursts or fires.

METHANE PREDICTION METHODS USED IN POLISH MINES

In most Polish mines methane hazard is predicted by the methane release forecasting. The forecasts become the starting point to determine the methane control strategies as well as the ventilation system in the face region and the methane drainage system. Output capacities are also evaluated based on the projected methane emission levels.

Methane capacity forecasts for the face regions are of key importance for safe longwall operations in the condition of methane hazard, underlying the designs of the air supply to the face region. They are also required to define the methane control measures through the selection and location of ancillary ventilation equipment and methane flow meters and sensors. When the predicted methane capacity is high enough to preclude safe longwall operations in the given ventilation conditions, the forecast is also used to design the adequate degasification scheme.

Forecasting methods in widespread use in Poland include that launched in 1978 following the regulation by the Ministry of Mining (MG) (Frycz, Kozłowski 1979), that developed by B. Kozłowski in the experimental mine Barbara (KD method) (Frycz, Kozłowski 1979). Nowadays a new dynamic method of absolute methane capacity forecasts is becoming popular, developed by E. Krause and K. Łukowicz (Krause, Lukowicz 2000).

In accordance with the MG method, the longwall region is expressed as the sum of methane fluxes released from the output mined from the longwall face and the seams surrounding the coal seam being mined. The KD method takes into account the methane flux released from the longwall face and tailgates. The dynamic method takes account of methane fluxes from the output mined on the longwall face, from the surrounding seams, from surrounding sandstone strata and other major fluxes from the longwall region.

Prediction methods outlined here differ in the number of methane fluxes they are able to handle and in the applied flux calculation procedures. That is why predicted results may differ considerably.

The common approach in all these methods is the averaging of methane fluxes released during the mining operations over the specified face range length by averaging the parameters used as input data to the forecasts to be made for the rock strata over and below the considered section.

APPLICATION OF NEURAL NETWORKS TO THE DESCRIPTION OF METHANE RELEASE TO MINE WORKINGS

Mining operations lead to the release of methane to active workings as well as abandoned workings. The process of methane release to mine workings is most complex; it is affected by a number of geological and mining-related factors. The occurring pressure gradient indicates the gas movement towards the mine workings, where pressure is lower than gas pressure in the rock strata. Pressure distribution inside the rock strata is treated as a spatial and unsteady process (Roszkowski, Szlązak 1999; Szlązak 2000; Szlązak 2005). The process of methane release to the mine workings is governed by a nonlinear complex function. Relationships derived from nonlinear equations are most complicated and so far no analytical methods are available to yield the general solution. In such cases approximate mathematical models are employed, which are then adapted in the course of the learning process. Plant identification in neural networks involves the model construction and finding model parameters such that the model's responses to the applied excitations should be equal, within the admitted tolerance levels. That is why the methane hazard evaluation method is proposed that makes use of neural networks.

In many cases predicted and real methane emission levels vary considerably and a neural network might easily become a verification tool (Borowski, Szlązak 2004). During the mining operations methane is released to the face region from the coal seam being mined as well as coal seams below and above, disturbed by longwall operations. Methane emissions come from:

- coal body (longwall panel or face zone); methane emissions directly to the air mixing zone in the mine working, no matter what the ventilation conditions;
- caved-in zone (abandoned workings); methane emissions to the mixing zone in the mine working, indirect, through the filtering zone, dependent on the airing conditions;
- mined coal, methane release from coal being hauled.

The adopted model of methane emission to the longwall region is shown in Figure 3.

The prediction method of methane release to longwall faces utilises all data available after longwall operations are over. Measurements were taken in five gassy face zones with the methane drainage systems provided. Longwall zones were ventilated by the U-configuration systems, the retreating mining method was selected with caving-in. Observations were recorded everyday over the whole face range, yielding 300 data records.

The measurement method consisted in simultaneous measurements of airflow velocity and methane concentrations. These parameters were registered by sensors being a part of a methane metering system. Each day the measured quantity was plotted on the graph, the weighted average was obtained and methane concentration computed to derive the amounts of air in the longwall zone and the amount of methane removed through the ventilation system. Measurements were taken by a sensor located near the outlet from the longwall zone.

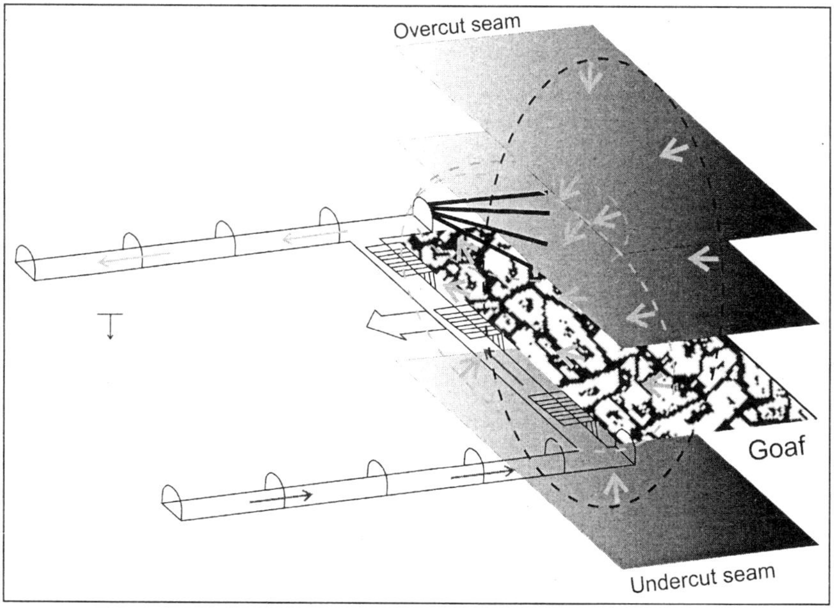

Figure 3. Model of methane emission to mine workings

Readouts were taken of the following methane drainage parameters: the amounts of captured methane, methane concentration, depression (pressure drop) in the drainage network, the number of active drainage holes. Variations in the recorded parameters were calculated per shift and per day. Daily output was obtained having taken account of face geometry.

The face range, coals seam thickness and inclination, mining depth and the face length were read out from mine maps. Methane capacity of the coal seam being mined and of seams disturbed by longwall operations are obtained from methane capacity measurements. The types of rock to be found in the roof and floor strata are determined on the basis of rock profile analysis.

An assumption is made that the desorption zones where coal seams are disturbed by longwall operations should cover the strata up to 150 m above and down to 50 m below the coal seam being mined. For rock strata disturbed by longwall operations the weighted average equivalent methane bearing capacity is computed (separately for undercut and overcut seams). The weight factor expresses the distance between the coal seam being mined and that disturbed by mining operations. Equivalent methane bearing capacity of rock strata disturbed by mining operations is obtained from the formula:

$$M_{zn,zp} = \frac{\sum M_i \cdot h_i \cdot \left(1 - \frac{h_i}{h_{p,n}}\right)}{\sum h_i} \tag{1}$$

where: M_{zn} – equivalent methane bearing capacity of undercut seams, m^3/Mg_{csw}; M_{zp} – equivalent methane bearing capacity of overcut seams, m^3/Mg_{csw}; M_i – methane bearing capacity of the i-th seam disturbed by mining operations, m^3/Mg_{csw}; h_i – distance between the coal seam being mined and that disturbed by mining operations, m; h_p – extent of the desorption zone in the undercut seam, h_p = 150 m; h_n – extent of the desorption zone in the overcut seam, h_p = 150 m.

The method allows for computing the amounts of released methane over 100-metre sections of the face range.

Table 2 summarises the data used in methane emission predictions. All relevant data are characterised in terms of arithmetic average, standard deviation, minimum and maximum value. The maximum and minimum values used in the learning process determine the applicability range of the method. Detailed calculations are provided elsewhere (Szlązak et al. 2008).

Table 2. Input and output data used in learning of the neural network (Szlązak et al. 2008)

Specification	Arithmetic mean	Standard deviation	Minimum	Maximum
Range of face, m	243,5798	154,3261	0,9	572,5
Coal production from the face, Mg/day	3213,1	1413,0	150,2	7690,8
Seam thickness, m	2,419	0,493	1,75	3,0
Face length, m	193,7	25,84	135	220
Inclination, degree	9,10	2,82	3,0	12,0
Methane bearing capacity, m^3/Mg_{csw}	5,379	0,879	3,70	7,32
Face depth, mppm	-573,6	98,27	-700	-440
Equivalent methane bearing capacity of undercut strata, m^3/Mg_{csw}	0,5177	0,471	0,2458	1,397
Equivalent methane bearing capacity of overcut strata, m^3/Mg_{csw}	1,2027	0,6789	0,4380	2,034
Amount of air, m^3/min	1660	402,6	548,7	3135,5
Pressure drop in methane drainage system, mm Hg	112,96	28,68	0	150
Concentration of methane drained, %	44,23	11,35	0	60
Methane removed through the ventilation system, m^3/min	9,88	3,68	1,73	20,92
Methane captured in the recovery system, m^3/min	7,23	2,97	0,00	14,20

Table 3 compiles regressive characteristics of neural network models (for the learning, validating and test subsets):

- average value of the output variable;
- standard deviation computed for the output variable;
- average error of the output variable (expressed as a difference between the preset value and that obtained on the model's output);
- average absolute error of the output variable (expressed as the absolute value of the difference between the predicted and real output value);
- standard deviation of errors for the output variable;
- correlation factor expressed as the standard R-Pearson's correlation factor for real values and those obtained on the model's output.

The ultimate row shows the R-Person's correlation factor between the predicted and real output data. For the predicted amounts of vented methane and that captured in the drainage system the correlation factor for the learning set is 78% and 90%. For the validation and testing subsets these coefficients have lower values yet remain within the acceptable limits. The required accuracy levels in predictions should not be less than 65% and the applied model ensures that accuracy level. The model affords us the means to determine the amounts of methane released to the longwall zone and that captured by the methane drainage system.

Methane emissions to mine workings are compiled in Figure 4, providing a comparison between the measurement and calculation data.

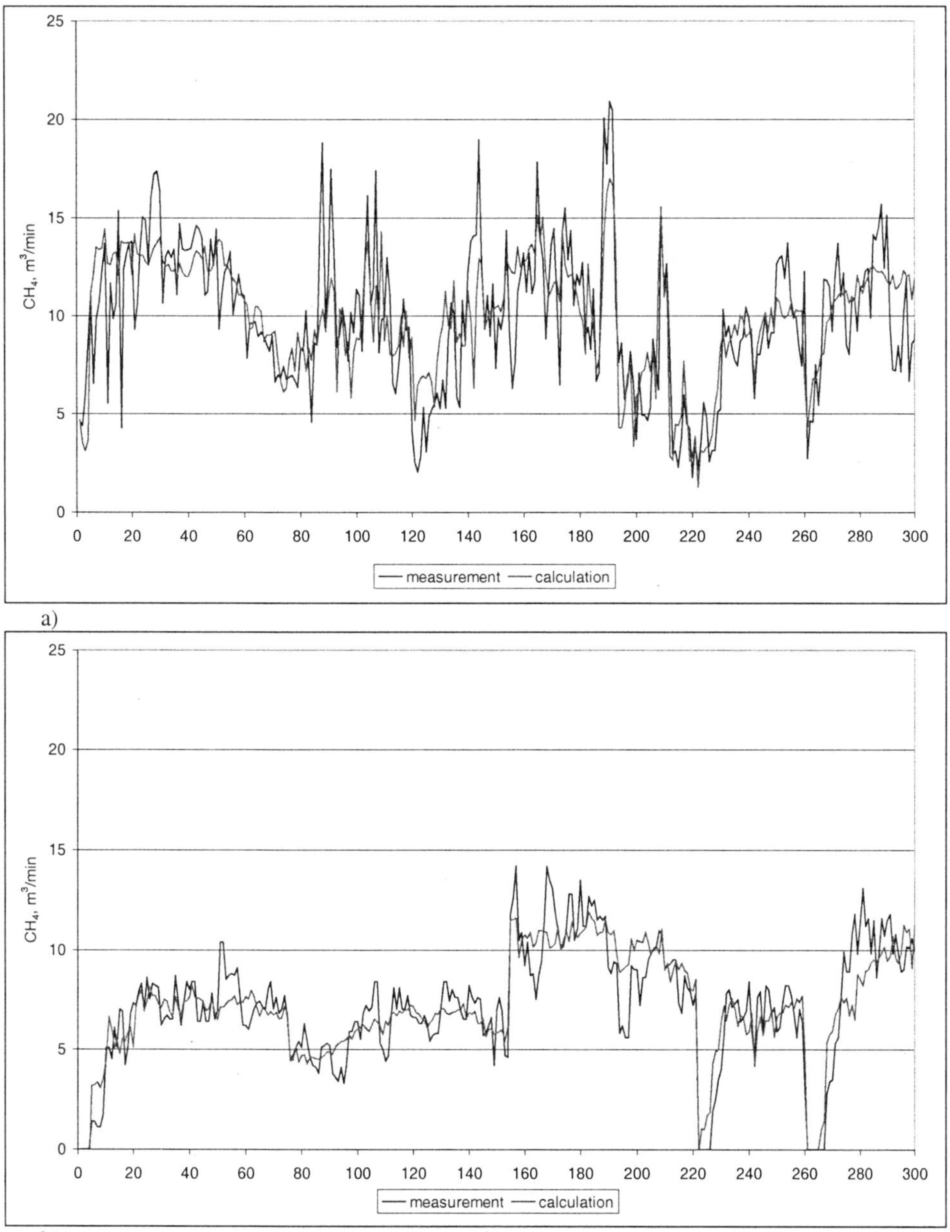

b)

Figure 4. Measurement and calculation data (with the use of neural networks):
a) vented methane, b) methane captured by the drainage system (Szlązak et al. 2008)

Table 3. Statistics of the neural network learning process (Szlązak et al. 2008)

Specification	Learning subset		Validation subset		Test subset	
	Vented methane, m^3/min	Methane captured in the drainage system, m^3/min	Vented methane, m^3/min	Methane captured in the drainage system, m^3/min	Vented methane, m^3/min	Methane captured in the drainage system, m^3/min
Mean value	10,36	7,34	9,93	7,70	8,84	6,53
Standard deviation	3,64	2,87	3,66	2,91	3,53	3,05
Standard error	-0,07	0,06	0,33	-0,26	0,31	-0,05
Error deviation	2,26	1,25	2,78	1,49	2,59	1,50
Average absolute error	1,75	0,98	2,20	1,17	1,92	1,15
Correlation factor	0,78	0,90	0,66	0,86	0,70	0,87

CONCLUSIONS

1. In the approach used to date methane contributions to the mine air are treated as a function of strata disturbed by mining operations and related to geological conditions and the mining method.
2. The analysis of methods used to determine the amounts of methane released to mine workings reveals that there are several major determinants of this process. The knowledge of these factors is not readily available, at least with the use of present methods.
3. Empirical models have been heavily criticised for major simplifications and the use of few variables in complicated calculations that are involved in methane emission forecasts.
4. Simulation models are used to capture gas flows in the coal seam and hence are widely applied in design and evaluation of methane drainage systems before the mining operations are commenced. These models emulate the real life conditions to a large extent, though they display a high degree of complexity.
5. Commercial software, is now available too, though its applications in mines are limited as it does not afford us the means to handle methane emissions from different sources: coalbed methane, strata disturbed by mining activities and from gobs (abandoned workings).
6. Application of the CFD approach does not provide any technical solutions required to determine methane emission levels; it only helps in evaluation of methane hazard.
7. Short-term forecasts can hardly be classified as methane predictions; they should be rather treated as observations of mining operations and methane release to the mine workings. The principal aim of these observations is to identify the potential methane hazard in the given working, taking into account the face advance rate and the daily output from the face, by comparing the predicted and real values.
8. Computer program might provide some assistance in searching for the optimal solution to the methane hazard problem and finding the best methane control plan. In order to solve those complex problems, it is required that the methane hazard evaluation system should be developed to identify other potential hazards: rock bursts or fires that might occur beside methane emissions.
9. Prediction models in the future should be more universal and less empirical than the existing ones and their complexity should not raise any difficulty in ongoing forecasting of methane emissions.
10. The methane hazard prediction and evaluation method has been proposed that makes use of neural networks, enabling us to determine the amounts of methane released to the mine workings and captured by the drainage systems.

REFERENCES

[1] Airey E.M. 1971: A Theory of Gas Emission in Coal Mining Operations. Presented at the 14th International Conference of Directors of Safety in Mines Research, Donezk, USSR.

[2] Badura H. 2001: Zastosowanie teorii szeregów czasowych do prognozy krótkoterminowej metanowości. Zeszyty Naukowe Politechniki Śląskiej, Górnictwo, z. 250.

[3] Barker-Read G.R., Radchenko S.A. 1989: Methane Emission from Coal and Associated Strata Samples. International Journal of Mining and Geological Engineering, Vol. 7, No. 2, pp. 101–126.

[4] Borowski M., Szlązak N. 2004: Prognozowanie wydzielania metanu do wyrobisk ścianowych w kopalniach węgla kamiennego. Materiały Szkoły Eksploatacji Podziemnej, Szczyrk, 23–27 lutego.

[5] CEC 1988: Investigation of Firedamp and Its Emissions in Coal Seams. Final Report on ECSC Contract 7220-AC/819, Report EUR 11477EN.

[6] CEC 1989: Short-Term Forecasting of Methane Emission Levels Using Continuously Monitored Data. Final Report on ECSC Contract 7220-AC/832, EUR 12326EN.

[7] Cybulski W., Myszor H.: 1974: Próba określenie ogólnej zależności wydzielania metanu od wielkości wydobycia. Archiwum Górnictwa, T. XIX, z. 1.

[8] Dixon D.W., Longson I. 1993: A Statistical Method for Methane Prediction and Improved Environmental Control. Proceedings of the 6th US Mine Ventilation Symposium, Salt Lake City, UT, USA (21–23 June).

[9] Durucan S., Daltaban T.S., Shi J.Q., Foley L. 1993: Permeability Characterisation for Modelling Methane Flow in Coal Seams. Paper 9315, presented at the 1993 International Coalbed Symposium, Tuscaloosa, AL, USA (17–21 May).

[10] Dziurzyński W., Krach A., Pałka T. 2002: Metoda wyznaczenia rozkładu stężenia metanu w sieci wentylacyjnej z uwzględnieniem systemu monitoringu. Materiały konferencyjne Szkoły Aerologii Górniczej, nr 2.

[11] Frycz A., Kozłowski B. 1979: Przewietrzanie kopalń metanowych. Wyd. Śląsk, Katowice.

[12] Jensen B., Gillies A.D.S., Anderson J.M., Jones N., 1992: Review of Methane Emission and Prediction Research in Longwall Coal Mines. In The Australian Institute of Mining and Metallurgy Proceedings, No. 1, pp. 11–17.

[13] Krause E., Łukowicz K. 2000: Dynamiczna prognoza metanowości bezwzględnej ścian. (Poradnik techniczny), GIG, KD „Barbara", Katowice, Mikołów.

[14] Lunarzewski L.W. 1992a: Gas Prediction, Recovery, Control and Utilization Applicable to the Mining Industry. Presented at the Workshop on Recovery and End Use of Coalbed Methane, Katowice, Poland (16–21 March).

[15] Lunarzewski L.W. 1992b: The Role of Predicted and Measured Gas Emission in Coal Mine. Gas Control. Presented at the Symposium on Coalbed Methane Research and Development in Australia, Townsville, Australia.

[16] Lunarzewski L.W., Lunarzewski A.L., Pilcher R.C. 1995a: A New Approach to Predict Underground Gassiness for Design of Gas Capture and Ventilation. Proceedings of the 7th US Mine Ventilation Symposium, pp. 61–66.

[17] Moll A.T.J., Lowndes I.S. 1994: The Computer Simulation of the Optional Design and Performance of Methane Drainage Ranges. Mining Engineer (London), 154 (394).

[18] Noack K., Opahle M. 1992: Progress Achieved in the Prediction of Gas Emission. Proceedings of the 5th International Mine Ventilation Congress, Johannesburg, South Africa, pp. 79–87.

[19] Patton S.B., Fan H., Novak T., Johnson P.W., Sanford R.L. 1994: Simulator for Degasification, Methane Emission Prediction and Mine Ventilation. In Mining Engineering (Littleton, Colorado), Vol. 46, part 4, pp. 341–345.

[20] Roszkowski J., Szlązak N. 1999: Wybrane problemy odmetanowania kopalń węgla kamiennego. Uczelniane Wydawnictwa Naukowo-Dydaktyczne, Kraków.

[21] Saghafi A., Jeger C., Tauziede C., Williams R.A. 1987: A New Computer Simulation of In-Seam Gas Flow and Its Application to Gas Emission Prediction and Gas Drainage. Presented at 22nd Conference Internationale des Institutes de Recherches sur la Securite dans les mines, Beijing, China, 4–6 November.

[22] Saghafi A. 1989: A Study into Face Gas Emissions during Heading Development. Presented at the 23rd International Conference of Safety in Mines Research Institutes, Washington, DC, USA, 11–15 September.

[23] Szlązak J. 2000: Przepływ powietrza przez strefę zawału w świetle badań teoretycznych i eksperymentalnych. Rozprawy Monografie, Akademia Górniczo-Hutnicza im. Stanisława Staszica w Krakowie, Uczelniane Wydawnictwa Naukowo-Dydaktyczne.
[24] Szlązak N., Szlązak J. 2005: Filtracja powietrza przez zroby ścian zawałowych w kopalniach węgla kamiennego. Akademia Górniczo-Hutnicza, Uczelniane Wydawnictwa Naukowo-Dydaktyczne, Kraków.
[25] Szlązak N., Szlązak J., Borowski M., Obracaj D. 2008: Ocena stanu zagrożenia metanowego i temperaturowego w rejonie ścian eksploatacyjnych. Akademia Górniczo-Hutnicza, Uczelniane Wydawnictwa Naukowo-Dydaktyczne, Kraków.
[26] Tauziede C., Pokryszka Z. 1993: Dynamic Prediction of Methane Emission at Longwalls. The 25th International Conference on Safety in Mines Research Institutes, Pretoria.

International Mining Forum 2009, Sobczyk, Kicki & Saługa (eds) © 2009 Taylor & Francis Group, London, UK, ISBN 978-0-415-80428-8

The Exploitation Depth and Bump Hazard in the Mines of the Upper Silesian Coal Basin

Mirosława Bukowska
Central Mining Institute, Katowice, Poland
e-mail: m.bukowska@gig.katowice.pl

ABSTRACT: Bump hazard is conditioned by three groups of factors. One of them is a group of natural factors (geological). They result from a diastrofic and sedimental development of a carboniferous coalfield of the Upper Silesian Coal Basin. Bump hazard in the USCB has changed over the years – both in the aspect of number of bumps as well as their intensity. The above mentioned results from changing geological and mine conditions, therein most of all from the permanent increase of the exploitation depth (currently about 8 meters per year) and the necessity to exploit the residual parts of deposits. It results also from motor requirements of coal mines which have intensified the exploitation. This subsequently is conductive to the intensity of the dangerous geodynamic occurrences in a rock mass. Over the last years the number of bumps in coal mines of the Upper Silesian Coal basin has decreased significantly as a result, among other things, of the exploitation technology improvement as well as bump prevention. Currently the highest number of bumps occurs in the range of depth between 700 and 770 meters what results from the intensity of coal exploitation at this range of depth. The highest number of bumps is still registered in the group of beds 500. The depth of exploitation, as a factor indirectly influencing the bump hazard, is considered in the methods of the assessment of rock mass susceptibility to bumps. The example is the geological and geomechanical method of the assessment of rock mass susceptibility to bumps GEO (Bukowska red. 2009).

KEYWORDS: Mining, geomechanics, rock mass, depth, bump, bumps hazard

INTRODUCTION

Bump hazard relates to mining of various mineral resources exploited with the underground system. There are also recognized cases of geodynamic occurrences, described as bumps, in the opencast marble mine. Bump hazard is one of the main natural hazards in the underground mining- both coal and ore. Its occurrence results directly from the natural properties of rocks and rock mass and is conditioned by mine works conducting. Although occasionally bump hazard may occur in a bed not influenced by exploitation events or in bed deposited in close neighbourhood and not disturbed tectonically (Konopko, Patyńska 2008).

The bump is a specific case of a seismic tremor. It results in the diminution of the mine working utility or an interruption of its utility and the safety of its utilization as well as danger for the staff. Bumps cause frequently the occurrence of other hazards, for example methane or fire etc. Each bump is accompanied by a seismic tremor with adequately high seismic energy. Simultaneously only about 1% of seismic occurrences, with their energies that may cause a bump, results in a bump

(Dubiński, Konopko 1995). Currently there can be specified two types of seismity in the Upper Silesian Coal Basin: mine and mine-tectonic. The character of seismity and the mechanism of bumps generation change together with change of exploitation depth. Kidybiński (2003) divided bumps considering the mechanism of their generation in the groups as follows:

1) bed bumps – generated by the transgression of a piece of rock mass strength in the zone of a large stress concentration;
2) bed bumps – initiated by the seismic tremor, however being mainly the result of a large stress concentration in the bed;
3) tremor bumps – damages in mine workings caused mainly by the energy and location of a seismic tremor focus.

The above-mentioned author has advanced a hypothesis that as the exploitation depth increases in the mines of the USCB, the bumps change their character from bed bumps to tremor ones.

BUMPS IN THE UPPER SILESIAN COAL BASIN

USCB is characterized by a complex structure. Initially it had been a piedmont subsidence basin but in further stages of its development it evolved into a mountainous subsistence basin (Kotas 1995). The subsistence basin has developed in the Precambrian massif (Caledonian-Variscan structural horizon of bed). It is filled with carboniferous productive formations (Variscan-Upper-Silesian molasse horizon). The youngest structural horizon in the USCB is built by cover formations of a productive carboniferous. Most of tectonic structures in the USCB originated in a compressible stress field or with a participation of a horizontal stress component, connected with the Variscan orogeny. During the Alpine orogeny the new tectonic structures arose or the old ones were rejuvenated. Some of the authors, on the basis of research performed in mine workings, point also at the occurrence of the area of substructure subsidential or epeirogenic vertical moves during forming of tectonic structures (Goszcz 1980). In the structure of the USCB one may distinguish the area of folding tectonics (the west part of the basin), the area of folding-block tectonics (the north and north-west part of the basin) and the area of disjunctive tectonics (the remaining and the largest part of the basin).

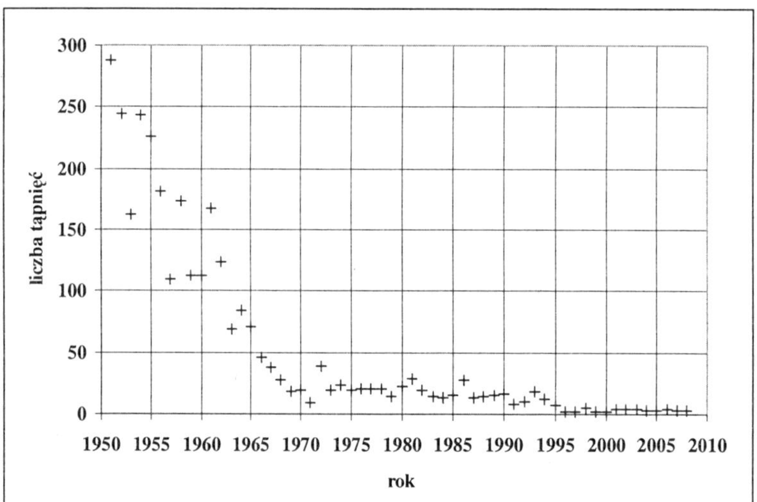

Figure 1. The number of bumps in the years 1951–2008 in the USCB (Patyńska 2007)

The issue of bumps occurrence in the USCB has existed almost from the beginning of mining in this area. In the turn of XIX and XX century mining had already been strongly developed in the Upper Silesia and tremors had become an occurrence strongly noticeable on the surface. The beginning of the XX century (years 1909–1912) was the period when tremors resulted in bumps in the mines of Upper Silesia.

The oldest description of bumps in the USCB dates back from 1858. It relates to the occurrences that had taken place in the "Fanny" mine. Initially, the term "bump" had been understood as "exfoliation of a mine working wall, coal spattering, which was accompanied by various acoustic phenomena" (Gustek, Kociela 1986). Together with the development of Polish coal mining the appearance of bumps has intensified in the USCB. In the turn of years the 1940–1950 a few hundred bumps per year were recorded (the maximal number of bumps was 506 in the year 1949). The development of exploitation technology as well as gradually improved methods of bump prevention caused a systematic decrease of number of bumps despite the stable increase of coal exploitation until the end of '70s of XX century and increase of exploitation depth. In the above-mentioned period there were noticed a few dozen of bumps per year. Since the '90s of XX century a significant decrease of number of bumps in coal mines of the USCB has been noticed. Currently their number reaches a few bumps per year (Fig. 1).

The significant decrease of number of bumps in the USCB is an effect of:

- the decrease of number of active mines;
- the decrease of an amount of output (lower number of active roadway walls);
- an improvement of exploitation technology;
- a modification of an attitude to the assessment of rock mass susceptibility to bumps (Bukowska 2002, 2003, 2005, 2006);
- more effective bump prevention;
- an implementation of the latest achievements of science and technology by the managements of mines;
- an activity of mine supervision in the scope of implementing of science achievements to by-law of Geological and Mine Law.

THE FACTORS DETERMINING BUMP HAZARD

Bump hazard is determined by various factors. The most important are natural reasons (geological), mine reasons (technical), and organization ones (Tab. 1). Bump hazard shall not be identified with a rock mass, coal and barren rocks susceptibility to bumps, which is one of the natural reasons of the occurrence of this geodynamic phenomenon.

Table 1. The factors determining bump hazard

Natural = geologic factors	Mining = technical factors	Organizational factors
An exploitation depth	A system of exploitation and roof management	Mistakes in works performance
Geomechanical properties and geologic history of a rock mass	An excessive concentration of stress as a result of exploitation events	Mistakes in hazard prevention
Lithology and petrography of strata, tectonics	Concentration of mining works	Too little investments in mines
Thickness of a coal bed	The parameters of the front of exploitation (length, height, velocity of progress)	Lacks in staff training
Watering		

It should be emphasized that one of the main factors determining the occurrence of bump is the depth of constantly increasing exploitation. Also an increasing depth of exploitation deteriorates the natural conditions of a rock mass in the aspect of its susceptibility to bumps and bump hazard.

AN INFLUENCE OF DEPTH ON BUMP HAZARD OF A CARBONIFEROUS ROCK MASS

While analyzing an influence of various factors on bump hazard the two aspects of the issue ought to be distinguished: a potential bump hazard resulting mainly from bed and surrounding rock mass properties as well as an actual hazard resulting from the conditioning of conducted exploitation. A few aspects should be entertained while considering a depth of bed from the point of its influence on rock mass susceptibility to bumps and bump hazard. Some of them are as follows:

1. in the USCB exploited coal beds are numbered from 100 (the youngest) to 700 (the oldest). The exploitation of beds 800 and 900 was finished at the end of XX century. The reason of closure of mines exploiting beds 800 and 900 was their non-profitability and deposits depletion. A complicated structure of the basin (lithostratygraphy, continuous and non-continuous tectonic) causes that the secular chronology of coal beds does not stand for an increase of depth of their occurrence. Taking into consideration a full lithostratygraphic profile of the USCB one may notice that generally numbering of beds increases with their depth. However none of groups of beds can be attributed a constant range of depth of bed occurring as it differs in various part of the USCB and mining areas of mines.
2. The changes of stress state in an original rock mass and in a rock mass disturbed with exploitation.
3. The changes in the quality of rock mass represented in diversification of physical parameters value, including geomechanical ones.
4. The changes in a degree of carboniferous rock mass watering and diversification of hydrogeological parameters values.

Among the factors conditioning the occurrence of bump the most important is structure of a rock mass understood as a configuration of strata with various properties including the presence of layers susceptible to tremors and exploitation events as well as large concentration of mining works conducted in the given bed. The above mentioned factors are connected with each other and have particularly unfavourable influence at the exploitation depths extending 600 m. They determine then the actual (real) bump hazard. The negative influence of main factors determining the bump hazard grows with increasing depth. Simultaneously the higher energy of a seismic tremor and the deeper, the smaller number of factors may become the reason of bump occurrence. The mining practice has shown that while at least four factors occur at the same time, there have been noticed bumps with tremor energy of 10^3 J. While tremor energy exceeds 10^7 J only one factor may evoke a bump (Konopko, Patyńska 2008).

The depth of exploitation and an original stress state in a rock mass resulting from it indirectly influence on rock mass susceptibility to bumps. The depth of coal beds occurrence is a result of the geological history of a carboniferous rock mass. The coal beds in the USCB are currently recognized to the depth of 1200 m.

The stress state significantly influences on the origination of geodynamic occurrences in a rock mass, and ipso facto on generating natural hazards including bump hazard. The stress state in a rock mass is shaped by original stress (gravitational), caused by the weight of roof layers, and tectonic stress (resulting from tectonic processes taking place in the earth's crust). Another group is constituted by exploitative stress resulting from the influence of mining activity. The state of stress in a rock mass changes with the depth and is modified by exploitation at given depth. A layout of stress in a rock mass resulting from deeper and deeper mining exploitation is becoming more disadvantageous because of bump hazard.

Observations and research conducted in mines of the USCB during the exploitation indicate that in the carboniferous rock mass, at relatively small distances, are often found significant changes of geological and mine conditions. These changes concern especially geomechanical properties of rocks which mostly decide about the occurrence of natural hazards and difficulties in mining processes. A very important factor influencing on forming of rocks properties was a tectonic stress. Fields of tectonic stress arose as a result of an activity of vertical moves of deep bedrock blocks as well as horizontal strain. The tectonics of the USCB is thought to be complicated. A composite tectonics indicates on very large values of tectonic stress which has caused a formation of various structures such as anticlines and synclines as well as large faults (most of them normal). The complexity of geological structure of the USCB expresses also in a diversification of geomechanical properties of coal of each stratigraphic group and barren rocks. A regional occurrence of bumps in the USCB results from a diverse structure of the basin and is tightly connected with distofic and sedimential development of the region. A variability of geological structure, including lithology and petrographic formation in the vertical profile as well as in the length of layers, reflects in a number and intensity of bumps. Bumps in the USCB take place mostly in the main basin (especially its north wing), in the main saddle (dome of Zabrze, dome of Maczek, a south slope of the saddle), in the Bytom basin, kazimierzowska basin, jejkowicka basin (maily SW wing) (Bukowska red. 2009).

Coal beds are strongly diverse with regard to geomechanical properties which depend directly on depth. The most significant in their forming are processes of diagenesis which consist in conversion of loose sediment deposited in a sedimentation environment into a solid rock. The deeper the larger pressures are in the rock mass being the result of an overburden weight and thus of a layout of original stress occurring at the given depth.

The natural properties of rock mass, among other things a lithological formation, are significant factors in bumps occurrence. Most of researchers connect the bump occurrence with thick groups of sandstones accumulating large amount of elastic energy and having a possibility of its sudden release during fracture processes. This process is one of the main reasons of generating tremors of large energies. In the above mentioned process also presence of mudstones in a lithological profile of upper-carboniferous, which repeatedly demonstrate higher values of strength and elastic parameters in comparison to sandstones, is significant.

Table 2. An approximate participation of main types of rocks of upper carboniferous in the USCB (%)

Rocks	Layers 100 and 200	Layers 301–326 and 327–406	Layers 407–419 and 500	Layers 600, 700, 800 and 900
conglomerate	3	-	2	-
sandstones	81	27	53	47
mudstones	6	39	25	38
claystones	6	26	11	12
coal shales	< 1	< 1	< 1	< 1
coals	3	7	8	2
others	< 1	< 1	< 1	< 1

The properties of coal of given bed and layout consisting of rocks surrounding the bed determine a bump occurrence (Bukowska red. 2009). In mining practice there are recognized cases when, despite the occurrence of sandstone layers with their thickness of several to few tens of meters in the vertical profile as well as presence of seismic activity connected with their destruction, bumps did not occur during tens of years of exploitation. A participation of sandstones in the structure of the USCB is very diverse both in a vertical profile and alongside the length of layers. Among sandstone formations building the main substance of carboniferous formations in the USCB, sandstones prevail in structure of the upper Silesian sandstone series (beds 407–510). Banks of coarse-grained, medium-grained and fined-grained sandstones, locally conglomerated, among which coal beds with various thickness from thin to thick occur (bed 510 with its thickness up to 24 m), are the main substance of Upper Carboniferous in the USCB (Tab. 2) the higher number of bumps is connected to (Tab. 3).

Table 3. The number of bumps in the USCB in the years 1970–2008 in groups of beds (Patyńska 2007)

Number of bumps	Group of beds				
	200	400	500	600	700
	1	10	459	12	7

One of the most important geomechanical parameters of rocks analyzed in the aspect of rock mass susceptibility to bumps and bump hazard is uniaxial compression strength. A variability of an average compression strength of sandstones of the USCB for groups of beds with different age (100–800) is presented in the Figure 2. There was also developed a dependence of sandstones compression strength (layers potentially susceptible to tremors) on depth of their occurrence (Fig. 3) (Bukowska red. 2009). The dependence of uniaxial compression strength of coals of the USCB on depth of bed occurrence is presented in the Figure 4.

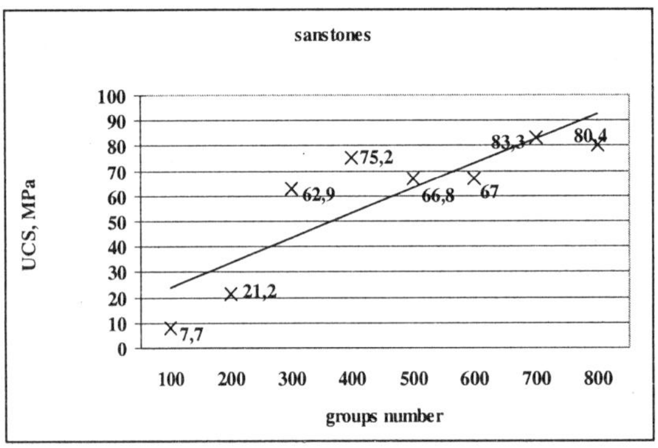

Figure 2. Variability of average carboniferous sandstones uniaxial compression strength in the USCB (Bukowska red. 2009)

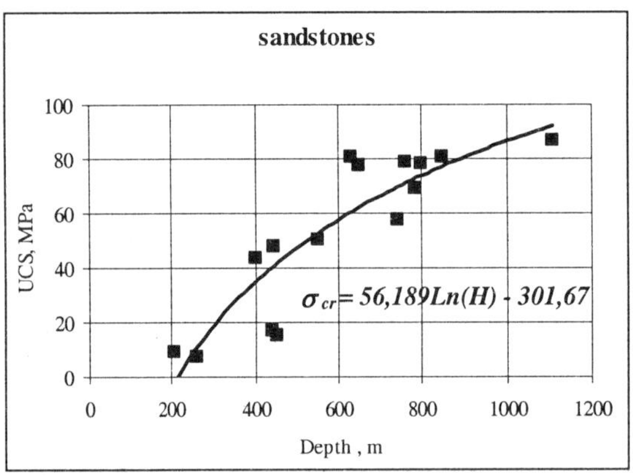

Figure 3. Dependence of sandstones compression strength on depth (Bukowska red. 2009)

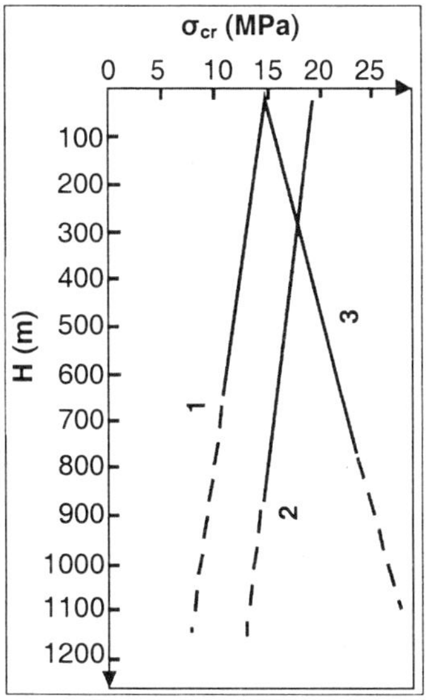

Figure 4. Changes of uniaxial compression strength of coals of the USCB with the depth of their occurrence (Kidybiński 1982): 1) bright coals $\sigma_{cr} = 14{,}5 - 0{,}006H$; 2) mean coals $\sigma_{cr} = 19{,}4 - 0{,}006H$; 3) mat coals $\sigma_{cr} = 14{,}2 + 0{,}0124H$

While considering the bump hazard, an important issue is the influence of water on rock mass susceptibility to bumps as a result of changes of some rocks parameters: hydrogeological (Bukowski 2002, 2007) and geomechanical (Bukowska, Bukowski 2005). In the hydrogeological profile of the upper carboniferous of the USCB there may be observed mainly fissure-porous water-bearing levels (Wilk red. 2003) built by sandstones and conglomerate. Sandstones of the upper-carboniferous are characterized by large variability of hydrogeological parameters. A very significant reduction of their water-bearing abilities and permeability may be observed with increasing depth (Kleczkowski et al. 1976). At large depths sandstones are characterized by very low permeability resulting from the processes of diagenesis of a rock mass. Generally there can also be noticed that the deeper the lower an effective porosity of sandstones is in the profile of water-bearing complexes (Fig. 5) whereas the changes of average permeability values with the change of depth do not present any regularity. In the conditions of the USCB there may be observed smaller recharges to deeper mining levels than to shallower ones (Fig. 6). The above mentioned is a result of fissures tightening in a rock mass and decrease of intergrains filtration coefficient value with an increase of exploitation depth.

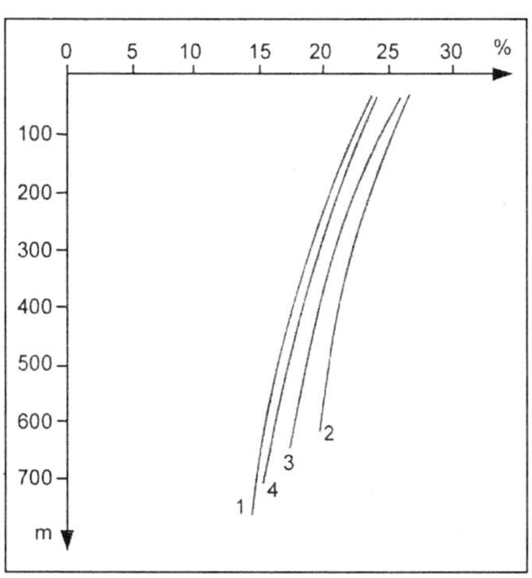

Figure 5. Diagram of tendency of porosity changes (%) of libiąskie and łaziskie layers with an increase of depth [m] by Wilk (1964); sandstones: 1 – coarse-grained, 2 – medium-grained, 3 – fine-grained, 4 – variously grained

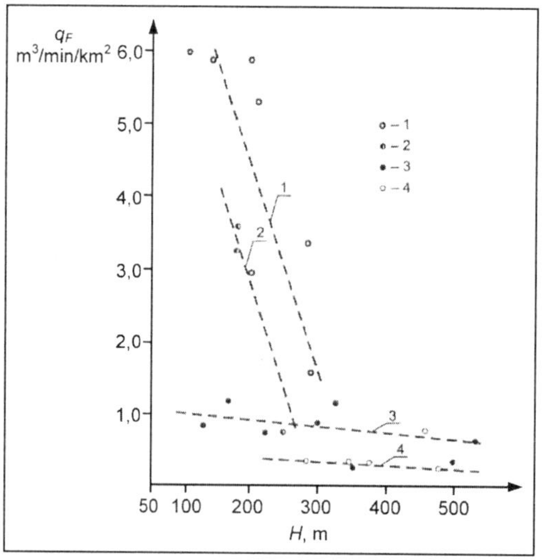

Figure 6. Dependence of unit recharges qF on depth H of mine in the USCB by Wilk (1965): 1 – cracovian sandstone series, 2 – upper Silesian sandstone series, 3 – mudstone series, 4 – paralic series

Laboratory research affirmed the weakening influence of water on geomechanical parameters, including uniaxial compression strength. It has been proved that sandstones lose on average 23–50% of their strength as a result of their watering, sandstones stratified with mudstones 16–53% and claystones lose about 50%. The least decrease of strength (1,5–5,5%) has been recorded for mudstones what is consistent with their little ability to absorb water considering their structure and texture (Bukowska red. 2009).

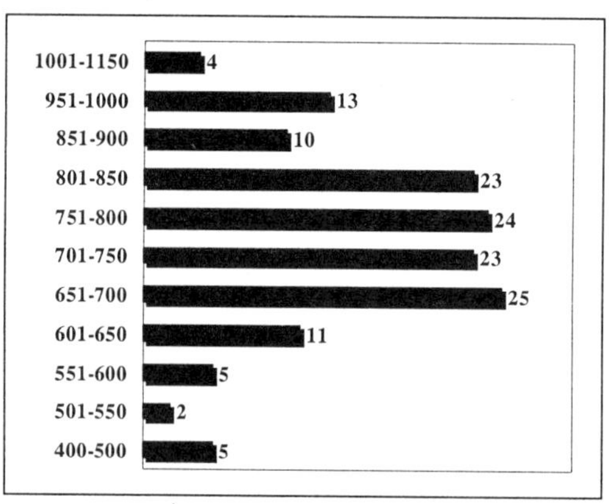

Figure 7. The depth of bumps occurrence in the USCB in the years 1987–2007

The changes of geomechanical and hydrogeological properties together with depth changes result in changes of geomecanical and hydrogeological coefficients values. The above mentioned has a direct influence on rock mass susceptibility to bumps and bump hazard in mine workings.

In the years 1987–1997 the minimal depth of exploitation connected with bumps increased from 400 to 620 m. In the years 1997–2007 there were 34 bumps during the exploitation of coal beds at the depth from 620–1150 m (Patyńska 2007), besides during this period there can be noticed no regularity in the layout of bumps in the USCB. It shall be noticed that the higher number of bumps relates to the depth 700–770 m – 12 bumps. It is undoubtedly connected with the depth of the current biggest exploitation in the USCB. During the exploitation of beds from groups 100–700 the highest number of bumps may be recorded in the group 500. In beds of groups 400, 600 and 700 during last ten years there have been noticed a few bumps in each. In more extended time, during last twenty years, the prevailing number of bumps has occurred while exploiting at the depth larger than 600 m. However because of an unfavourable influence of geological and mining factors bumps have occurred also at depths smaller than 500 m (Fig. 7).

The exploitation depth is taken into consideration in the methods of the assessment of rock mass susceptibility to bumps. One of them is the system of geological and geomechnical assessment of rock mass susceptibility to bumps (Bukowska 2005, Bukowska red. 2009). The system considers seven natural factors influencing on the susceptibility to bumps. One of these factors is the depth of coal bed layout. In the GEO classification the depths larger than 700 m are scored with the highest number of points.

SUMMARY

The bump hazard is situated among natural hazards and occurs in the rock mass susceptible to bumps. Rock mass susceptibility to bumps results from natural factors (geological). These factors have been formed as an effect of a diastrofic and sedimentation development of a carboniferous basin in the USCB. The main natural factors determining the susceptibility to bumps and bump hazard are:

- an exploitation depth,
- thickness of an exploited bed,
- lithology,
- tectonics of a rock mass,
- geomechnical properties,
- rock mass watering.

Among natural hazards occurring in the underground mining, the influence of exploitation depth is particularly demonstrated in the case of bump hazard.

Together with the depth increase, geomechanical and hydrogeological properties of rocks as well as geomechanical and hydrogeological conditions of rock mass change. Moreover also the character of bumps changes with the increase of exploitation depth – bed bumps give up their place to tremor ones.

The acquisition of coal in the new parts of a rock mass, constant deterioration of exploitation conditions (exploitation within pillars and remnants of the bed, under-level exploitation), may cause the increase of frequency of bumps occurrence in the USCB in the future and thus an increase of bump hazard.

Currently in the USCB with exploitation up to the depth of 1200 m the largest exploitation is conducted at the depth of 700–770 m and in this range of depths the highest number of bumps is noticed.

A constant change of exploitation conditions resulting from the increase of depth of coal exploitation in the USCB (currently about 8 meters per year) may cause in the future that bumps in the coal mining within the USCB will be more intensive than it is nowadays – probably it may cause the occurrence of mining catastrophes.

REFERENCES

[1] Bukowska M. 2002: Geomechanical Properties of Rocks from the Rockburst Hazard Point View. Archives of Mining Sciences, Vol. 47, No. 2, pp. 111–138.

[2] Bukowska M. 2003: Post-Critical Properties of Carboniferous Rocks in the Upper Silesian Coal Basin under Varying Strain Rate and Confining Pressure Conditions. Archives of Mining Sciences, Vol. 48, No. 4, pp. 561–575.

[3] Bukowska M. 2005: Mechanical Properties of Carboniferous Rocks in the Upper Silesian Coal Basin under Uniaxial and Triaxial Compression Tests. Journal of Mining Science, No. 2, Vol. 41, pp. 129–133, Springer New York.

[4] Bukowska M. 2006: The Probability of Rockburst Occurrence in the Upper Silesian Coal Basin Area Dependent on Natural Mining Conditions. Journal of Mining Science, No. 6, Vol. 42, pp. 570–577, Springer New York.

[5] Bukowska M. (red) 2009: Kompleksowa metoda oceny skłonności do tąpań górotworu w Górnośląskim Zagłębiu Węglowym. Wyd. GIG, Katowice.

[6] Bukowska M., Bukowski P. 2005: Uwagi dotyczące wpływu zawodnienia na ocenę skłonności górotworu do tąpań dla obszarów granicznych kopalń czynnych i zlikwidowanych w GZW. Prace Naukowe Instytutu Geotechniki i Hydrotechniki Politechniki Wrocławskiej 75, Seria Konf. 41, str. 49–58.

[7] Bukowski P. 2002: The Water Storage Capacity of a Carboniferous Rock Mass and its Impact on the Flooding Process of Mine Workings in the Upper Silesian Coal Basin. Archives of Mining Sciences, No. 47, 3, pp. 385–412. Warszawa – Kraków, Wyd. PWN.

[8] Bukowski P. 2007: Zagrożenia wodne w kopalniach węgla kamiennego w Górnośląskim Zagłębiu Węglowym w dobie restrukturyzacji górnictwa. Górnictwo i Geoinżynieria. Kwartalnik AGH, Rok 31, Zeszyt 3/1, str. 81–91.

[9] Dubiński J., Konopko W. 1995: Kierunki zwiększania efektywności profilaktyki tąpaniowej. Przegląd Górniczy, Nr 4, str. 21–27.

[10] Goszcz A. 1980: Wpływ napięć tektonicznych na niektóre własności skał i warunki górnicze w północno-wschodniej części Górnośląskiego Zagłębia Węglowego. Zesz. Nauk. AGH, Geologia, Nr 27, str. 9–92.

[11] Gustek M., Kociela W. 1986: Charakterystyka zagrożenia tąpaniami w kopalniach Górnośląskiego Zagłębia Węglowego. Kwartalnik AGH, Seria Górnictwo, Zeszyt 2.

[12] Kidybiński A. 1982: Podstawy geotechniki kopalnianej. Katowice, Wyd. Śląsk.

[13] Kidybiński A. 2003: Zagrożenie tąpaniami w górnictwie światowym – rozpoznanie i zapobieganie. Prace Naukowe GIG, Górnictwo i Środowisko, Nr 1, str. 5–35.

[14] Kleczkowski A.S., Motyka J., Wilk Z., Witczak S. 1976: Zmiany niektórych cech hydrogeologicznych skał zwięzłych z głębokością. Przegl. Geol., T. 24, Nr 5, str. 263–268.

[15] Konopko W., Patyńska R. 2008: Warunki występowania tąpnięć w kopalniach węgla kamiennego. Przegląd Górniczy, Nr 1, 2008, str. 12–17, Katowice.

[16] Kotas A. 1995: Górnośląskie Zagłębie Węglowe. W: The Carboniferous System in Poland. Zdanowski A., Żakowa H. (red.). Prace PIG, CXLVIII, Warszawa, Państ. Inst. Geol., pp. 124–134.

[17] Patyńska R. 2007: Zagrożenie tąpaniami. [W] Konopko (red.). Raport roczny o stanie podstawowych zagrożeń naturalnych i technicznych w górnictwie węgla kamiennego. GIG, Katowice.

[18] Wilk Z. 1964: Relation between Porosity and Depth of Sandstones of the Łaziska and Libiąż Beds (Upper Carboniferous). Upper Silesian Coal Basin. Bull. Ac. Pol. Sc., Ser. Sc. Geol. Geogr., Vol. 12, No. 2, pp. 113–118.

[19] Wilk Z. (red.) 2003: Hydrogeologia polskich złóż kopalin i problemy wodne górnictwa. AGH, Uczelnie Wydawnictwa Naukowo-Dydaktyczne, Kraków.

International Mining Forum 2009, Sobczyk, Kicki & Saługa (eds) © 2009 Taylor & Francis Group, London, UK, ISBN 978-0-415-80428-8

Possibility of Application of the Mechanized Coal Excavation in the Mines with Underground Exploitation in Republic of Serbia

Miodrag Denić
JP za PEU, Resavica, Republic of Serbia
e-mail: miodrag.denic@yahoo.com

Zlatko Dragosavljević
JP za PEU, Resavica, Republic of Serbia

Slobodan Kokerić
JP za PEU, RMU "Soko" Sokobanja, Republic of Serbia
e-mail: skokeric@sezampro. rs

ABSTRACT: The primary goal of underground coal mining is the increase of production. Room-pillars methods are used for coal excavation in the mines with underground coal exploitation in Serbia today, so if the increase of production is to be achieved it is necessary to introduce mechanized excavation in mines in which mining-geological conditions allow it, using long-wall methods of excavation.

In some of our coal mines at the proper time there was used a long-wall method of excavation (RMU "Rembas" – Resavica, RMU "Bogovina" – Bogovina), so there are certain experiences which are very significant for the future selection of MHP.

This paper includes the proposal of introducing the mechanized excavation in the mines in which mining-geological circumstances allow it.

KEYWORDS: Exploitation, mechanized excavation, shaft, long-wall, coal

INTRODUCTION

The possibility of the increase of using one's own resources of the solid energetic fuel in the process of solving the energetic crisis which is dominant in our region is a true challenge for the mining as a science. The well-known fact is that the biggest resources of energetic resources in Serbia are solid fossil fuels, and the safe energetic resource is only the one which comes from one's own energetic resources.

Our economy policy didn't have the right approach to solving the global energetic crisis which provoked many-years recessions of the developing countries. With the first worlwide energetic crisis (1965), at the moment when the coal industry reaches its peak in its production development, the country leaves the former serious care about energetics, and just when in 1964 reaches its peak in coal production by underground exploitation of about 3.850.000 t.

Reorientation to energetic fuels (oil and gas) imported, in the next ten years cost Serbia the loss of production capacities and physical destruction of the greatest number of mines with underground coal exploitation.The rest of mines lose, until today, its developing component, together with the keeping up the pace with the contemporary and modern production achievements in the coal industry, transferring to the struggle for the bare existence as a burden for society and one's own standard, because the total coal production from underground exploitation reduced to 600.000t at the beginning of 21st century on the annual level. The complexity of the exploitation conditions, new market conditions of the economy and the more difficult realization of profit led underground exploitation to a very difficult state whose further existence will depend on its readiness to introduction of new technologies in all phases of coal exploitation.

Present active mines with underground exploitation in Serbia are characterized by relatively complex geological-mining conditions which had or can have a negative impact when using the mechanized coal excavation.

The analysis of the results for the different types of MHP for the long-wall excavation used until now in underground exploitation is given in this paper, and for the shafts of mines "Rembas" and "Bogovina" as well as the proposal for the solution for mechanized excavation in mine "Soko".

ACOMPLISHED RESULTS AND PROBLEMS WHEN WORKING WITH MHP IN UNDERGROUND COAL EXPLOITATION IN SERBIA

Pit "Shaft 10–West field" RMU "Bogovina"

In shaft "Pit 10–west field", the excavation of coal was executed with the use of MHP Hemscheidt type 3200-20/30, combine-machine KBH, front conveyor Grot and collecting conveyor GTS-572, with the use of *sub-level* method, in period of 1978–1990.

During the exploitation there were no serious holdups except for the regular maintainance of the equipment, with the next parameters:

- total length of the front wall was 60 m with 40 sections installed;
- progress 40–45 m/month, and
- annual production capacity 200.000 t of commercial coal.

The accomplished results of work for the mentioned period are shown in Table 1.

Table 1

Working year	t.e.c./year	t.c.c./year	t.c.c./month	t.c.c./day	m/month
1978	121.146	84.802	7.066,83	282,64	42,6
1979	257.679	180.375,3	15.029,77	601,16	48,6
1980	258.459	180.921,3	15.076,75	603,7	47,6
1981	259.758	181.830,6	15.069,16	602,7	45,6
1982	259.576	181.703,2	15.059,77	601,9	46,6
1983	256.007	179.204,9	14.933,6	597,3	44,6
1984	260.754	182.578,8	15.214,8	608,5	45,3
1985	259.758	181.830,6	15.152,5	606,4	47,1
1986	258.735	181.114,5	15.092,8	603,7	46,3
1987	259.345	181.541,5	15.129,25	605,1	48,3
1988	260.126	182.088,2	15.174,07	606,9	45,4
1989	259.438	181.606,6	15.133,8	605,3	47,6
1990	137.759	96.431,3	12.053,8	482,12	41,6

The work of MHP Bennes-Marrel in shaft "Jelovac" of mine "Rembas"

After an unsuccessful use of (1972 and 1977/1978) of hydraulic pillars Salzgitter with the upper beam (rafter), which gave the efficiency 16 t/wages, in 1979 in shaft "Jelovac" there was an installation carried out and MHP Bennes-Marrel MB 17-28S began to work which was in use until 1982. Problems during the work with this type of MHP reflected in the following:

- in 1979, out of 214 working days 37 days were lost or 17,3% and:
  - damage on excavation machine-combine lasting 4 days;
  - damage of the front supporting conveyor lasting 13 days, and
  - holdups because of external factors lasting 20 days.
- Average shift holdups lasting 210 minutes on MHP (for the observed period 1979–1980) because of:
  - damage on excavation machine-combine lasting 65 minutes;
  - damage of the front supporting conveyor lasting 68 minutes;
  - damage on collecting conveyor lasting 28 minutes;
  - insurance of demolished and unstable parts of the front wall lasting 38 minutes;
  - individual damages on sections lasting 8 minutes;
  - damages on energetic installations lasting 1 minute, and
  - other holdups lasting 2 minutes.

Results of work that is the achieved efficiency per years with the given MHP are shown in the Table 2.

Table 2

Working year	t/month	t/day	m/day	t/wages
1979	12.146	511	1,261	10,44
1980	12.194	486	1,23	9,69
1981	4.137	170	0,51	3,26
1982	12.209	540	1,26	9,75

MHP Hemscheidt-ESO in shaft "Jelovac" in "Rembas" mine

In "Jelovac" mine, the excavation of coal with MHP Hemscheidt-ESO type 3200-20/30 and combine-machine Eickhoff EW 170-L, front conveyor Eickhoff EKF-3 and collecting conveyor GTS-572, was carried out in the period from 1986–1988. During a year out of total 350 achieved working days, there were holdups lasting 45 days or 12,8%, and:

- holdups in work because of damages on excavation machine-combine lasting 60,33 working hours;
- holdups in transportation system lasting 716,98 working hours, and
- other planned and unplanned holdups lasting 296,2 working hours.

Work on a long-wall excavation went at suitable exploitation conditions.Total average strength of the seam from floor to roof with waste rock intrusions was 9,12 m. Average apparent fall of the coal seam in direction of the front wall progressing was 2–5°,and in direction of front wall line 3–9°. The results of work with this MHP are shown in Table 3.

Table 3

Working year	t/month	t/day	m/day	t/wages
1986	13.850	551,7	1,09	16,50
1987	12.938	509,5	0,90	15,70
1988	13.250	504,3	0,97	15,87

MHP Hemscheidt in shaft "Senjski rudnik" in "Rembas" mine

Coal excavation in shaft "Senjski Rudnik" in period from 1987 to 1991, was carried out with blind long-wall method (for the first time in the mining practice present at the time being in the reagion of the Republic of Serbia) with the use of MHP Heimscheidt 30/28 with total 17 sections, length 28 m, excavation machine Eickhoff-Essa 150-L with one cutter of installed power 150 KW and collecting conveyor GTS-572 of the motor power 2×55 KW. The average achieved progressing was 29 to 35 m/ /month, with efficiency of 15 t/wages. The mentioned blind long-wall method worked without bigger holdups, as without measurable oxidation processes which in the previous excavation systems was a very common case.

In the excavation field, where the same thing was located, geological conditions were very suitable, which reflected on a successful and safe work of the emloyees and the equipment. The average strength of the excavation front was about 7 m; the average fall of the coal seam in the direction of the excavation front progressing was 3–5°, while in the direction of long-wall front line 14– –17°.

Except for the mentioned systems MHP, there were more attempt of mechanized coal excavation, in the Soko mine (half-mechanized long-wall type Salzgitter), in Aleksinac mines MHP type Vestvalia, and some more attempts, but all of them were unsuccessful in the essence of it.

THE PROPOSAL FOR THE SOLUTION OF THE MECHANIZED COAL EXCAVATION IN "SOKO" MINE

Mine designing and planning of the mining activities is a detailed and a comlpex process which demands an overall knowledge of the deposit geology, equipment characteristics and mining technology. During the designng, all stated factors are analized methodically as to get an idea solution, which would be the basis on which the different variants would be assessed.

The choice of the way of opening and the basic concept should be such that the period of opening that is the investing construction of mine be as less as possible, so that the regular production be established and the beginning of the return of the investments.

According to the available data and the geological interpretation of data obtained by research drilling from the terrain surface, research drilling from the shaft and data obtained by mining works, *the coal deposit of the east wing of the central field* of the "Soko" mine (Fig. 1) is characterized by a vertical depth of coal seam of 25 m, and horizontal of 30 to 40 m. Minimum width of the field is by the fault R-8 on K+120 m level and is 200 m, and maximum width is on K–20 m level and is about 700 m.

The coal seam in this limited field spreads from K+100 m to K–150 m, with the slope towards north from 35–45° and with the total balance of coal reserves of 16.730.898 t. Use of mechanized excavation, demands a total knowledge of tectonic characteristics and the position of the coal seam.

The choice of the excavation method is carried out on basis of a comlex analysis which gives important criteria as:

- the volume of capital investments;
- time of opening and reaching the full capacity, and
- the value of the production costs, and taking into account that for the long-wall method it is necessary to predict considerable means for the equipment, expensive works on opening and preparation of working terrain, but with very low production costs, while with the mechanized room-pillar method of excavation, the production of coal is achieved immediately, with less investments in equipment more production costs but with the lower degree of the used deposit.

The possible variants of the mechanized coal excavation in "Soko" mine can fall within the next:

- coal excavation by use of the Continuous Miner as an excavation machine in room-pillar excavations;
- coal excavation by use of the mechanized long-wall with one approaching room (the blind long-wall) and the overthrow of the coal deposit upper part, and
- coal excavation using the mechanized long-wall with two approaching rooms (circulating long-wall) and the overthrow of the coal deposit upper part.

In this paper the variant of the mechanized excavation for the mine "Soko" deposit conditions, *using the Continuous Miner as the excavation machine in room-pillar excavations.*

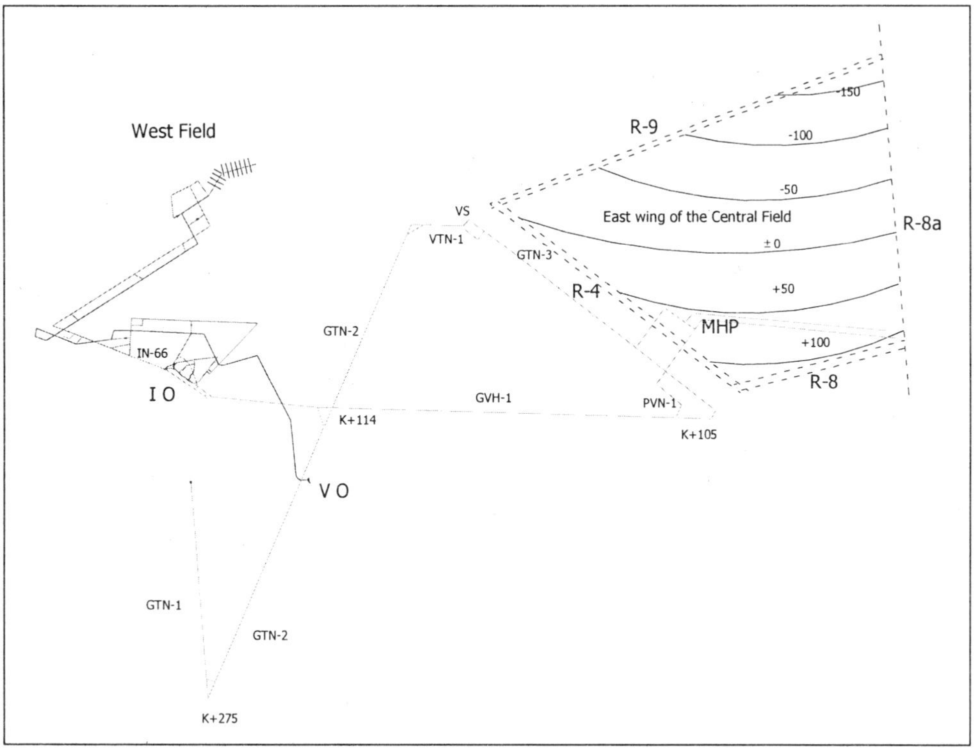

Figure 1. East wing of the central field of the shaft of "Soko" mine

Continuous Miner is a combined mining machine intended for excavation works, loading and transportation of coal on a small distance (coal reloading). This machine is already used over the world (mostly in the USA, Australia, Great Britain and South Africa Republic), as the excavation machine for the costruction of the underground halls.

Speaking of the use of the Continuous Miner as an excavation machine, with the regular choice and designing of some of the room-pillar methods, Continuous Miner can be applied with achieving the satisfactory efficiency. In this way the use of the long-wall methods and the significant investments related to them are-postponed.

Opening of the part of the excavation deposit with the Continuous miner, that is, a room-pillar method is executed by more entrances and transversal halls, which form the protection halls which can be excavated completely or partially during the withrawalor they are not excavated.

Deposit opening is carried out in this way out of reason:

- approaching the coal reserves with more paralel halls-entrances which are connected with each other by halls, constructed on an equal level-transversal halls, and
- construction of the protection pillars in a way that would enable their excavation (completely or partially).

Halls construction for purpose of approaching the coal reserves means a longer period of their use, which has a need for the supporting of a higher quality when the constructing mine halls, of minimum width, with enlarged pillar dimensions. The model for working out and the preparation of storeys with two paralel halls and transversal halls on each 20 m. Storey halls are classic transportation and ventilation ones, through which the free flowing and at constructing of mine halls and excavation, separate ventilation.

Halls dimensions of opening and working out of the excavation field are 4,5×4 m, and the supporting is carried out with the classic AT rockbolts in roof of the hall, while the side walls of the hall are supported with rockbolts made of plastic fabric which would be cut with Miner during pillar excavation. Because of the hall stability and safety of the workers and Continuous Miner, apart from AT rockbolts the halls are additionally insured with the steel frameworks of a rectangular shape on the axis distance of 1 m, which during the excavation and before cutting of Miner, are successively off.

Width of the pillars is 9 m, hall width is 4,5 m, and the hight of storey is 4 m the size of the possible hold of the Continuous Miner. Total width of the excavation front (horizontal width of the coal seam) is 36 m and contains of three pillars and two storey halls. Excavation would be done in segments of the length of 20 m, and always with the use of movable sections of hydraulic support. Generally, the hight of the storey is 5,0 m because there is a coal sheet put between storeys of 1 m width because of the stability, so that the works on a lower storey could be executed normally as for the working out and the preparation for the excavation.

Practically coal is obtained by the construction of halls and the exploitation of pillars without any coal extracting from the layer above the excavating level (horizontal concentration), and the previous work from the upper floors is demolished after the pillar exploitation and withdrawal of the sections of the movable mechanized support. The aim of the pillar excavation is to increase the recovery of the deposit which is during the very construction from 20 to 25%. Pillar excavation can be partial or complete. Partial excavation is carried out when because of unstable roof and dominant pressures the whole pillar cannot be excavated, that is Miner has no time to withdraw from the unsupported part of excavation before the roof has been demolished. Excavation of pillar parts is by rule carried out from entrance as it is shown in Figure 2.

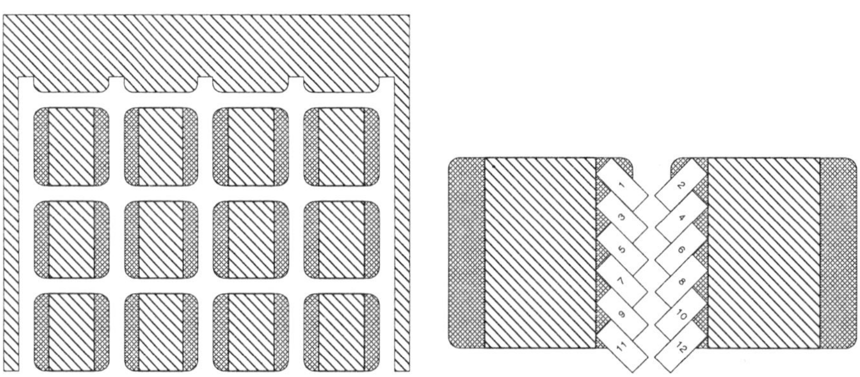

a) b)

Figure 2. Excavation of the pillar parts where as a temporary support movable sections are used

Double hatched pillar parts are excavatated evasively, in a way the mechanized hydraulic supports withdraw together with Miner. Placing the sections or pillars is carried out so to define the line of rift out of cutting zone, in order to unburden the roof in the next cut zone and give more time to Continuous miner.

Use of movable support gives an opportunity for a faster pillar exploitation by which the efficiency is increased and more important than that, Continuous miner and the workers withdraw faster from the endangered zone. In Figure 3 and Figure 4 two methods of the pillar excavation are shown by use of the mechanized support. What is characteristic for both methods is that the pillars are of the rectangular shape which defines the number of cuts that are needed for its excavation and width defines the lenth of each cut.

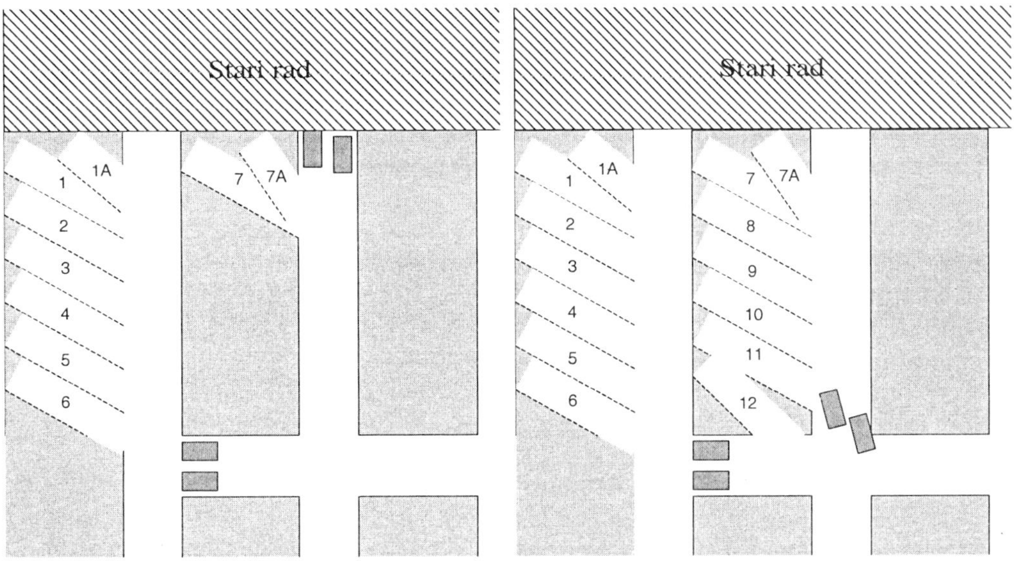

Figure 3. Pillar excavation by external cut using the movable mechanized support

Method which is shown on Figure 3 can be found in literature under the name "external cut", where Continuous miner from the zone with the stable roof (entrance) begins the pillar excavation with continual and paralel cuts, where the first cut is different only and its purpose is to create space for the positioning of movable sections as the space for Miner's entrance into the next cut under better conditions and greater use of l deposit.

Another method of excavation with the use of a movable support is known under the name Christmas tree (Christmas tree method) as shown in Figure 4.

Dynamics of works using such type of mechanized excavation is very flexible, work on getting the coal can be done in three production shifts, 25 working days in a month on average, and the rest five days are at disposal for other production activities so that the total month productivity is as follows:

$$Q_m = 216 \times 3 \times 25 = 16.250 \text{ t/month},$$

that is:

$$Q_g = 16.250 \times 12 = \mathit{195.000\ t/god}$$

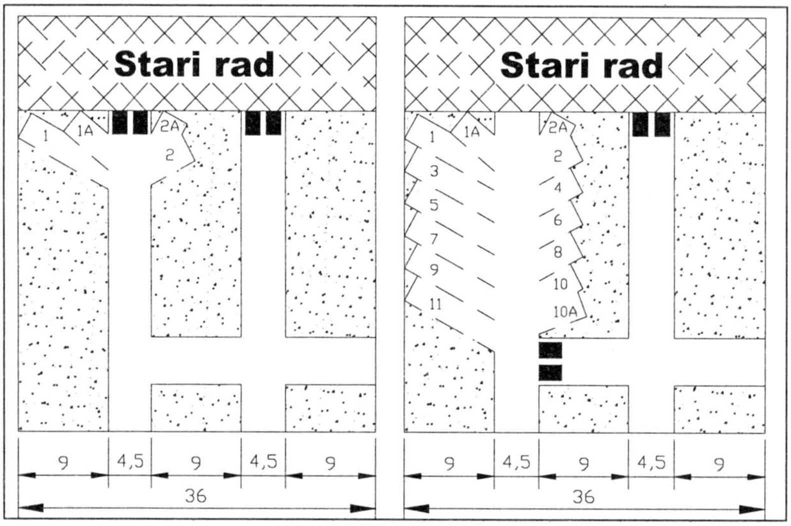

Figure 4. Pillar excavation with Christmass tree method

FINAL OBSERVATIONS

Improvement of technological coal exploitation process can be realized through next phases:
- the choice of the optimal model of the layer opening;
- use of modern methods of the construction and supporting of underground halls, and
- the right choice of the highly prodictive mechanization for the coal excavation.

Choosing and using the right solutions for the opening of deposits the conditions are created for the bigger production concentration, as in relation to the objects of opening, preparation and working out so as in relation to the basic production units.This can significantly influence the pre-conditions for using the contemporary technical-technological solutions and of course, for the increase of economizing in technological coal exploitation process.

Detailed and overall research and knowledge of the specific quality of deposit and working environment, true and reliable backgrounds necessary for designing, enables the choice of the most suitable model of mine opening, application of progressive and safe solutions of exploitation which increase the reliability of technological process greatly, the work safety and workers safety.

The basic concept and model of deposit opening should be such that the time of opening, that is, investment construction of mines be as shorter as possible, so to establish the planned system of production and the beginning of return of the investments.

Model for opening of "Soko" mine pit (phase I and phase II) represents a contemporary way of opening and working out the deposits with so called "floor" system. Opening is carried out with the system of the inclined transport ways from surface (k+397 m) and export shaft (k+170 m) and reaches k–44 m level, on which the first horizon of the "Soko" mine pit is located (Fig. 1).

The proposed model of the pit means that during the exploitation in the east field more horizons are formed, which will be functionally connected to the capital objects of pit opening. The chosen system leaves the possibility of a simple and rational subcutting of the remaining coal reserves in the north part of the SokoBanja basin under the fault R-10. Capital objects of opening are dimensioned so that ($P = 14{,}5\ m^2$) during the whole life of exploitation they enable functionality and sufficient porous ability, from the aspect of ventilation, pit servicing, production material delivery, shipment of coal and realization of other technical-technological demands.

Excavation dynamics dictates the need for a fast preparation of new excavation fronts that is the need for the increase of underground halls construction speed. In that sense one of the main problems which follows the underground coal exploitation in mines of Serbia is construction speed and stability of underground halls. In order to overcome this problem and the right choice of construction technology and supporting these halls, must begin with the realization on transfer technology whose main aim is the improvment of the status of the undeground halls that is the improvment of the quality of their construction, supporting and increase of their life duration, and that can be achieved with:

- mechanized mine hall construction, and
- adequate system of insurance of the underground halls.

Disintegration of the working environment, in order to create the free space because of underground hall construction, the natural balance of the working environment that was present until that then is worsened. Contemporary technological processes of disintegration of working environment find out the ways of steps which will as less as possible disturb the natural balance which means mechanized underground hall construction.

Basic criteria of the machine success for the mechanized underground hall construction are economy and reliability (availability) as well as the possibility of working in hard conditions characteristic for the underground coal exploitation. Under contemporary construction technology which enables the stability of underground halls and with it the economy (without reconstruction), which is only possible to achieve by mechanized underground hall construction, using combine with cutting head or Continuous Miner, with the adequate safety system as AT hanging support.

The most complex segment of the underground production system which in the essence defines the size of a mine is the system and the way of coal exploitation, that is the method of excavation. Research of methods and technologies of excavation represents research of possibility of using the excavation constructions and necessary installations and equipment for executing technological process of coal disintegration and then with the economic analysis the profitability or unprofitability of the suggested method is proved, that is, by use of economic parameters the optimal element value which can be changable in some spectrum of values is established.

The special significance represents the choice of complex equipment for its successful application. Basically, the task is to find the suitable type of equipment which could be suitable for the planned way of excavation in the conditions of the working environment in which it is used and from the versatile production programme of different types of complex equipment offered by the world market. The complexity of a successful solution of this task is more difficult because of the complex mining-geological conditions our coal layers are known for, and which are often so different that the acquire experiences in work on a mechanized coal exctracting from one later, can only be partialy used for another deposit.

Speaking of the Continuous Miner use as an axcavation machine, it can be said that with the increase of depth of carrying out the mining works and more difficult geological conditions, and when as an economically unjustified solution the use of long-wall methods of excavation is imposed, with the right choice and designing of some of the room-pillar methods, Continuous Miner can be applied with great successs with acheiving the satisfactory efficiency. In this way the use of long-wall methods and for them significant investments are postponed. Also, inspite the fact that the Continuous Miner is envisaged and designed as an excavation machine mainly for the coal seam excavation of thickness from 1,8 to 4 m, thanks to contemporary technical solutions and its construction improvments, the use of this machine is more significant for the excavation of seams of a bigger thickness.

The possible changes of angles of the main slope and even the changes of the floor characteristics can slightly change the characterictics of a mine, but also with small changes of a working method it is possible to continue the exploitation with this system without loss of productivity.

Broader introduction of methods for the mechanized excavation in coal deposits, with the possibility of a higher production concentration, with the possibility of adjusting, excavation mechanization of the changable structure characteristics of deposits and depths of excavations, opens the new possibilities of high concentration of production and increase of the specific capacity of the excavation. Conquering and introducing the new technology with the mechanized coal excavation, and particulary for the continual construction of underground halls for opening, basic and excavation preparation of a layer and the high degree of mechanization and automatization, ensures high speed of progress, shortenes the time of investment construction and enables the maintenance of the production capacity on high level, and in that way improves the total business economy.

REFERENCES

Čokorilo V., Milisavljević V.: Roof Monitoring Devices and Proposal for Introduction of AT Rockbolts in Colliery "Soko". Proceedings of ICAMC '98 – 13th International Conference on Automation in Mining, High Tatras, Slovačka, 1998.

Čokorilo V., Miljanović J., Bogdanović D., Denić M.: Razvoj podzemne eksploatacije uglja u svetu. Rudarski radovi, 1/2001.

Čokorilo V., Lilić N., Pinow C., Milisavljević V.: Mogući pravci modernizacije postojećih kapaciteta za proizvodnju uglja podzemnom eksploatacijom, sa aspekta značaja za energetski razvoj Srbije. Zbornik referata sa VII međunarodnog simpozijuma MAREN 2006, Beograd 2006.

Elaborat o rezervama uglja rudnika "Soko". JP za PEU, 2004.

Miličić M., Čokorilo V.: Mining Equipment Evaluation and Selection Based on Technical Characteristics. Proceedings of an International Conference: „Górnictwo zrównoważonego rozwoju", III Konferenja: "Energooszczędne i niezawodne maszyny górnicze", 30.11–1.12.2000, Gliwice, Poland, 2000.

Milisavljević V., Denić M., Tanasković M., Gašić V.: Mogućnost primene savremenih tehnologija izrade i podgrađivanja rudarskih prostorija u nekim rudnicima. JP za PEU, Zbornik radova sa VI međunarodnog simpozijum – MAREN 2002, Beograd 2002.

Denić M.: Doktorska disertacija: Analiza uslova za primenu visokoproduktivne otkopne mehanizacije za podzemnu eksploataciju strmih slojeva uglja velike debljine. RGF, Beograd 2007.

Milisavljević V., Denić M.: Savremene konstrukcije mašina za izradu podzemnih prostorija. Zbornik radova sa IV naučno-stručnog skupa: "Podzemna eksploatacija mineralnih sirovina u novim uslovima privređivanja", Beograd 2001.

Smart B., Čokorilo V.: The Importance of Using a Realistic Conceptual Model for Strata Control When Specifying Coalface Powered Supports. Proceedings of Conference with international participation: "Geomechanics Support of Mining Production", 3rd–7th June 1997, Nessebar, Bugarska.

Studija: Strategija razvoja energetike SRJ do 2020. Godine, sa vizijom do 2050. Godine, Energija, Godina II, Beograd 1997.

Studija: A Case Study and Analysis of Continuous Miners and Shuttle Cars on Steep Gradients. Smoky River Coal Company – Grande Cache, Alberta Canada.

Studija: Ispitivanje mogućnosti mehanizovane izrade i mehanizovanog podgrađivanja podzemnih prostorija ankerima sa dvokomponentnom smešom u rudnicima. JP za PEU, RGF, Beograd 2000.

Žugić M., Denić M.: Mogućnost primene savremenih tehnologija otkopavanja neregularnih slojeva uglja. Zbornik radova sa međunarodnog simpozijuma: "Korišćenje uglja u energetici", Ohrid 1997.

Đukić B., Đukanović D., Sanković Ć.: Ostvareni rezultati kod primene mehanizovanog otkopavanja u jamama RMU "Rembas" – Resavica. Zbornik radova VII Međunarodni simpozijum „Mehanizacija i automatizacija u rudarstvu i energetika", Beograd 2006.

Tehnička dokumentacija RMU "Bogovina" – Bogovina i RMU "Rembas" – Resavica.

International Mining Forum 2009, Sobczyk, Kicki & Saługa (eds) © 2009 Taylor & Francis Group, London, UK, ISBN 978-0-415-80428-8

The Influence of Heading Localization Depth on Their Stability

Stanisław Duży
Departament of Geomechanics, Underground Construction and Surface Safety Management, Faculty of Mining and Geology, Silesian University of Technology, Gliwice, Poland
e-mail: stanislaw.duzy@polsl.pl

ABSTRACT: The depth of heading localization is one of the main factors that decide on the maintenance conditions and their stability. The paper, on the basis of research results of rock resistance properties and their changeability in the depth of occurrence and the occurrence of changing stress in the rock strata, presents the problem of heading stability depending on the depth of occurrence. Having in mind the increase in heading depth localization, the changing state of stress and the variation of the surrounding massive resistance factors, using probabilistic methods, the influence of heading depth occurrence on its structure stability and reliability was specified.

KEYWORDS: Mining engineering, underground structure, heading stability

1. INTRODUCTION

Carrying mining exploitation causes exhaustion of natural resources, which for the maintenance of production level is connected to the necessity of exploitation accessibility and preparation of new deposits – often occurring on greater depth. Generally it is believed that the heading localization depth mainly influences on the occurrence of natural hazards and the worsening of heading stability maintenance conditions, which are caused by the increased values and changing stress distribution in the rock strata. The mentioned above unfavorable phenomenons, observed in headings localized on great depths, are dependent also on the values and changing of resistance and deformation factors of the rocks that build the massive.

To evaluate the heading stability, generally the limit states method is used. This method, in the procedures of building structure design, distinguishes the ultimate limit states and the serviceability limit states. In the currently used designing and heading support choice methods, the covering ultimate limit states method is used. With the increase of heading depth localization, causing the increase of many natural hazards, it is necessary to cover serviceability limit states methods, which makes it necessary to, among others, forecast heading convergence.

The basic task in the process of designing is the most precise definition of parameters that determine cooperation of the support with the rock strata to achieve the highest possible degree of structure reliability and safety.

2. BASIC RULES OF HEADING SUPPORT STRUCTURE RELIABILITY AND STABILITY

According to M. Chudek (Chudek 1986) the stability of a heading is defined by maintaining, in a certain time period, the needed dimensions of its cross-section and ensuring the safety of people, equipment and mining machinery from falling off walls and roof rock fragments and cavings. The support is stable when it does not shift entirely or partially under the influence of the pressure of the rock strata, which would threaten to destroy the support. The support structure stability is evaluated on the basis of the limit states method.

The stability of the whole structure or its parts, from the point of view of load capacity, should be expressed in reference to the limit states set, which separate desirable structure states from the undesired ones (PN-ISO 2394:2000). The following limit states can be distinguished:

- ultimate limit states – they suit the maximum load capacity or in certain cases the maximum permissible deformation or shift;
- serviceability limit states – they regard the normal structure usage.

The ultimate limit states include:

- partial or entire loss of structure balance treated as a stiff body;
- reaching the maximum load capacity in its sections, elements or joints through severance, or
- excessive deformation;
- transformation of a structure into a mechanism;
- reaching a state of instability of the structure or its part;
- sudden change of the accepted structural system into an other system.

Checking the support ultimate limit states, relies on showing that in every reliable structure section, for the design action the following condition is fulfilled (Duży 2007):

$$S_d \leq R_d \tag{2.1}$$

where: S_d – design value of action effect (general external force), R_d – design value of structure load capacity (structures element).

The serviceability limit states include:

- impermissible structure or foundation deformations,
- local structure damage,
- excessive structure vibration.

Checking the support serviceability limit states, relies on showing that the following condition is fulfilled for the design action:

$$E_d \leq C_d \tag{2.2}$$

where: E_d – action effect (i.e. stress, scratch width, shift, deflection, heading convergence), C_d – effect limit value.

To evaluate heading stability, taking input data changeability under consideration, probabilistic or halfprobabilistic methods can be used.

The evaluation of structure safety in the level II probabilistic solution is based on the assumption that the failure condition can be defined by this condition (Murzewski 1989):

$$Z_0 = P_0 - q_0 \geq 0 \tag{2.3}$$

where: Z_0 – safety margin, P_0 – support load capacity, q_0 – support load.

Two parameters of probability distribution, the expected value and the standard deviation, have decisive meaning. Structure safety is based on the “weakest link” idea.

The Cornell reliability coefficient t is accepted as the measure of safety:

$$t = \frac{\overline{P}_0 - \overline{q}_0}{\sqrt{s_{P_0}^2 + s_{q_0}^2}} \tag{2.4}$$

The value of the function of distribution of the p(t) reliability coefficient stands for the support structure safety probability, whereas the value [1 – p(t)] stands for the probability of structure failure (the supports' loss of stability).

In this method the conditions of reliability and safety of these types are used (Duży 2007):

$$p \le p_a \tag{2.5}$$

where: p – probability of stability loss, p_a – acceptable level of stability loss probability.

Depending on the desired class of heading stability maintenance conditions, the acceptable level of failure probability is:

- for good conditions – $p \le 0.03$,
- for satisfactory conditions – $0.03 < p \le 0.10$,
- for sufficient conditions – $0.10 < p \le 0.20$.

3. ROCK RESISTANCE AND DEFORMATION FACTORS AND THEIR CHANGEABILITY DEPENDING ON DEPTH OF OCCURRENCE

Research results of rock resistance and deformation properties and their changeability was presented in (Duży 2006 and 2007; Duży, Szafulera 2006 and 2007) among others. The analysis was carried out from the point of view of the following issues:

- the forming of rock compressive resistance that build the same layer on a grater area;
- the forming of rock compressive resistance that build the same layer that occurs various depth;
- the forming of rock compressive resistance depending on the marked position on the layers thickness;
- the forming of rock compressive resistance that build the various layers, which are included in various stratigraphy levels.

On the basis of the analysis of the gathered research data, one layer of sandstone, slime and clay, occurring on the whole area, in the roof were selected for further studies, and from the results sets measurements concerning these layers were separated.

Figure 3.1 presents an example of sandstone, occurring in the roofs coal stratum, compressive resistance distribution. From the presented distribution shows that in the roof of this stratum the sandstone reaches values from the range of 43.96–72.76 MPa. Having the compressive resistance in mind, it is characterized by relatively small changeability on the most of the analyzed area. A clear increase of resistance is forecast in the north-eastern region, the smallest increase is forecast in the south-eastern region.

Figure 3.2 presents the forming of a discovered, by in situ studies, compressive resistance of the analyzed rocks in individual layers, depending on the depth of occurrence. The presented charts show that there is no relation between the clay shale, which builds the analyzed layers, resistance and the depth of their occurrence.

Considerable variance of the rock, that build the same layer, resistance value causes large difficulties in assuming support design input data. Additional problems occur when heading on a lower level are designed, in the bounds of which there were no carried research or the research are located on higher levels. In that case, essential information can be the geotechnic parameter value changeability distribution, depending on the depth of occurrence. As the measure of rocks, which build individual layers, compressive resistance changeability, the coefficient that is a division off the standard deviation and the average value was accepted. The distribution of the compressive resistance changeability coefficient, depending on the depth of occurrence, is presented on Figure 3.3.

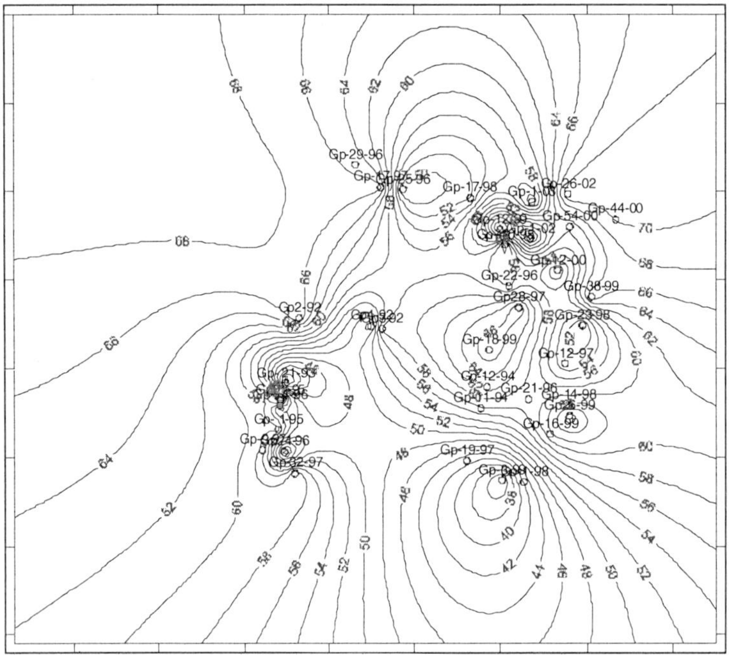

Figure 3.1. Distribution of rock compressive resistance value in analysed area (Duży 2007)

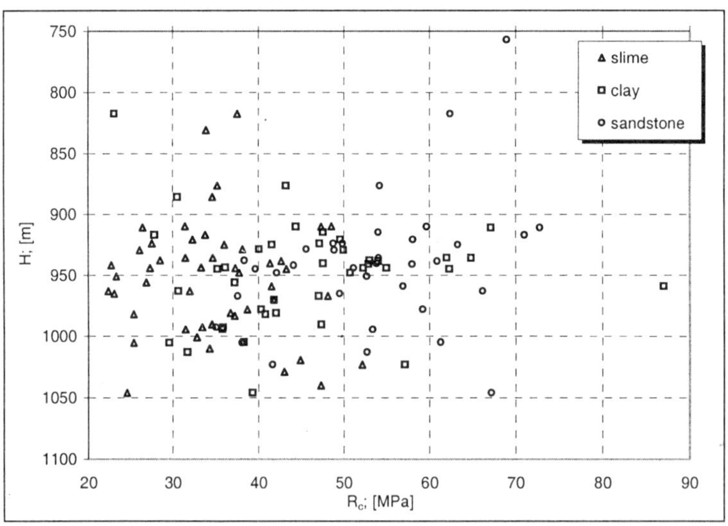

Figure 3.2. Distribution of rock compressive resistance value depending on depth (Duży 2007)

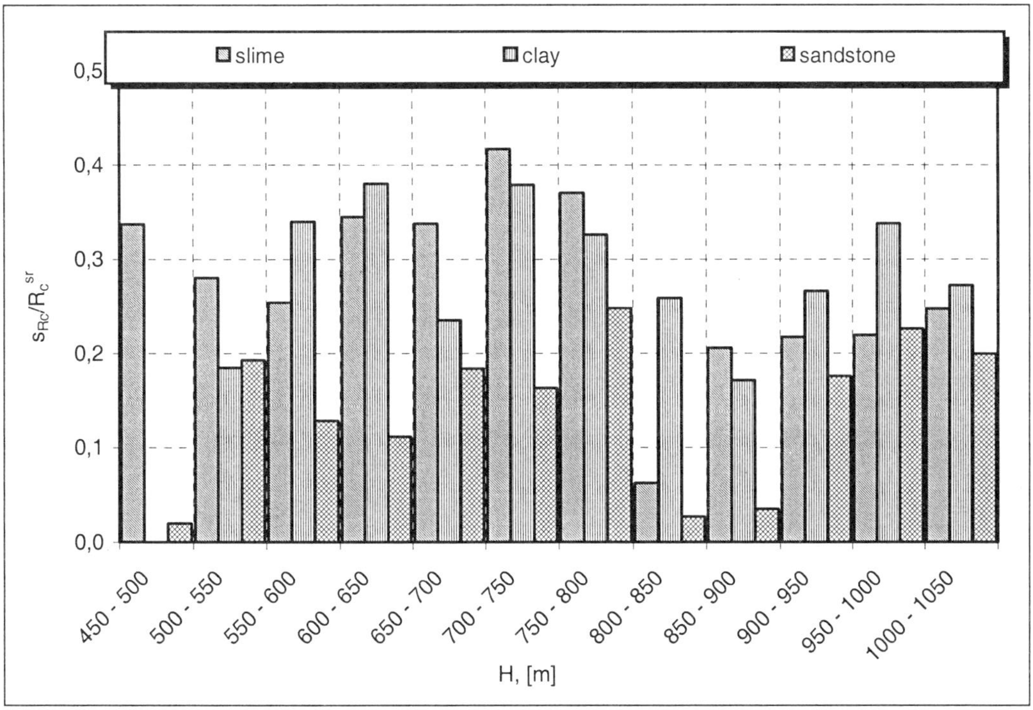

Figure 3.3. Distribution of rock compressive resistance changeability depending on depth (Duży 2007)

Having research results in the designing process, the problem of their interpretation occurs. Often the research results is statistically worked up, defining the average and variance values of individual geotechnic parameters. Compressive resistance is basic input data to design underground structures and often it is defined as one characteristic value for each layer (when using deterministic methods), or as a random variable of a determined probability distribution (when using probabilistic methods).

The presented above sandstone, slime and clay compressive resistance result analysis, though it only concerned one rock layer, it points out the complexity of the choice of input data process for design. The issue can significantly complicate when on the designed lower levels layers occur, that do not occur in the higher levels.

Except for the in situ research results, laboratory research, performed on the basis of 4 prospect holes that cover all of the of Orzesze layers depth of strata, results were analyzed. Figure 3.4 shows the forming of compressive resistance average values, defined for individual layers, depending on their depth of occurrence. The presented chart shows that slime occurring in the boundaries of the analyzed area is characterized with similar compressive resistance values and changeability, irrespectively of the depth of occurrence and the Orzesze layers localization in relation to the roof and floor. Clay and sandstone occurring in the boundaries of the analyzed area are also characterized by similar compressive resistance values and changeability, irrespectively of the depth of occurrence and the Orzesze layers localization in relation to the roof and floor.

Having in mind the rock strata stratigraphic structure and the related depth of occurrence of individual rocks, the hypothesis is often confirmed. Figure 3.5 shows the forming of sandstone compressive one-wall resistance average values, slime and clay occurring in the GZW, included into various stratigraphic levels (Duży 2006).

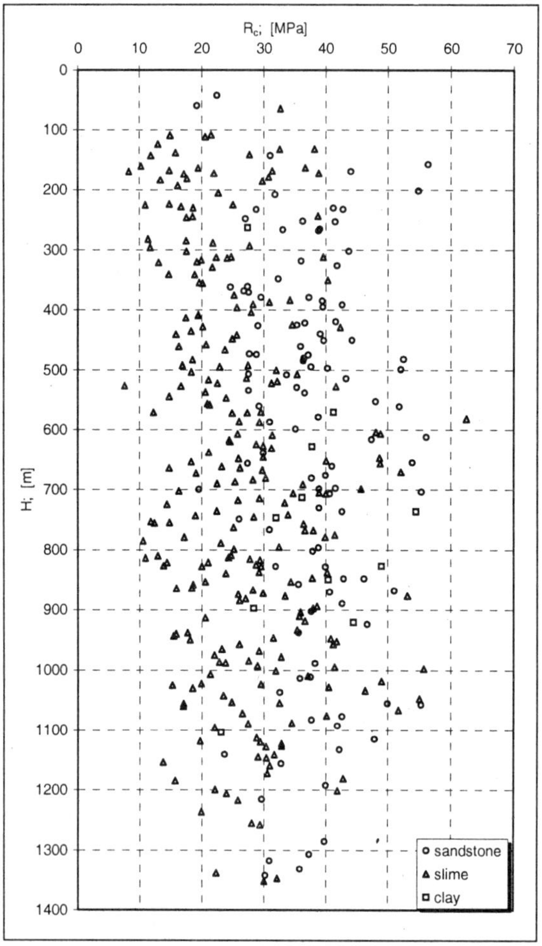

Figure 3.4. Distribution of rock compressive resistance value depending on clay, sandy shale and sandstone layer depth (Duży 2007)

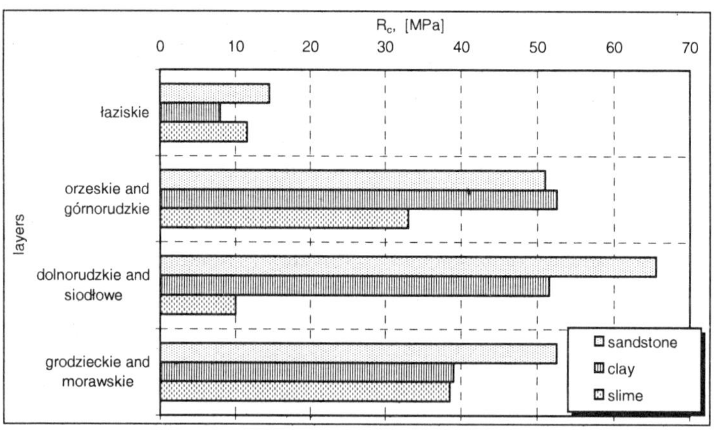

Figure 3.5. Diagram of rock compressive resistance, having in mind the stratigraphic level (Duży 2006)

On the basis of the carried analysis and theoretical considerations it can be stated that:

- rocks, in the boundaries of the heading characteristic design profile, deformation and resistance factors often adopt various values, even in the boundaries of the same layer, and the
- changeability along the heading coasting can exceed even 100%;
- rock strata deformation and resistance properties values, that are necessary to design underground structures, should be accepted having in mind their changeability;
- rock resistance values in a massive can be defined using statistically method;
- the distance of the analyzed heading section from the nearest information point has a significant influence on the level of accepted rock massive deformation and resistance properties values credibility;
- rock properties can change also in time, which can be caused by i.e. by driving in the region of the heading.

4. THE INFLUENCE OF DEPTH ON THE SIZE AND STRESS DISTRIBUTION IN THE ROCK STRATA

The other most important, decisive factor of heading stability maintenance conditions is the size and distribution of rock strata stress. The size of vertical stress in the intact rock strata on the H depth is generally accepted, and the horizontal component as a function of vertical stress and Poisson distribution. Taking the Poisson coefficient value of less then 1, the stress state vertical component is always greater then the horizontal component. The values of stress components increase proportionally to the increase of depth. In some calculative methods the vertical stress increases due to external action i.e. the influence of other headings, the influence of edges or seam remains. These factors often do not take under consideration the change of the relation between horizontal and vertical stress.

Carrying, in many regions of the world, research of rock strata stress state have shown that horizontal stress exceed vertical stress in most headings, that occur on depth of less then 1000 m. (Hoek, Brown 1978; Doaws, Hons 1992).

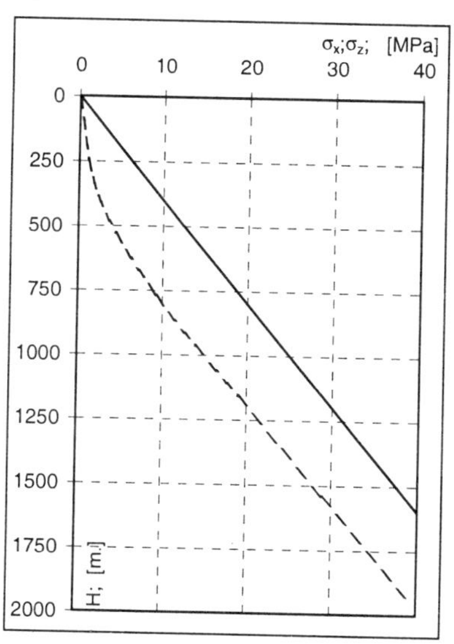

Figure 4.1. An example of vertical and horizontal stress chart, depending on depth (Chudek 2002)

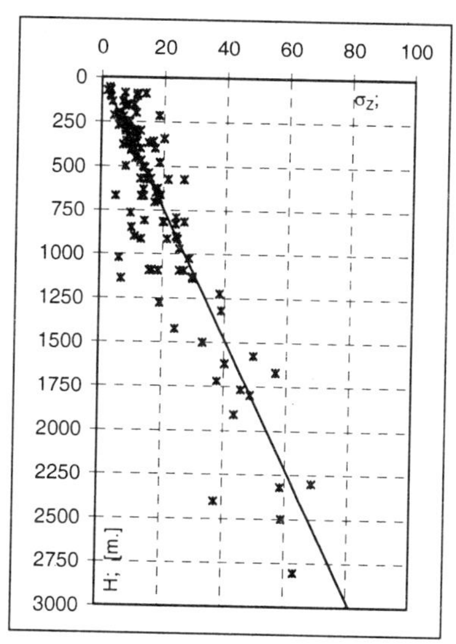

Figure 4.2. Vertical stress research results chart (Brown, Hoek 1978; Daws, Hons 1992)

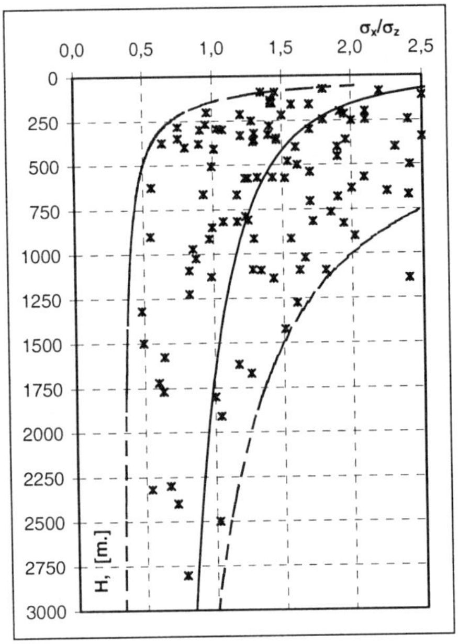

Figure 4.3. Horizontal stress research results chart (Brown, Hoek 1978; Daws, Hons 1992)

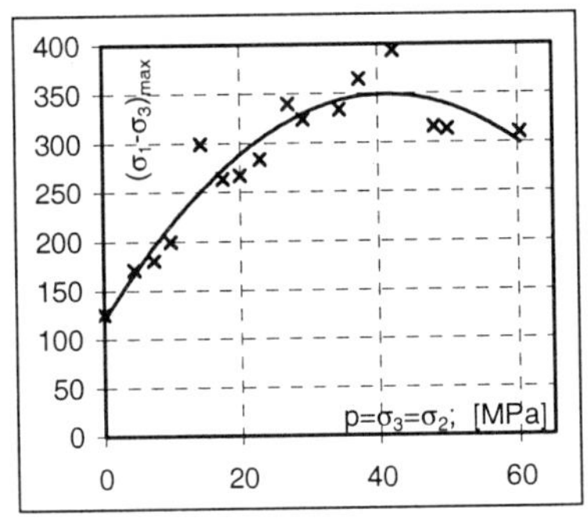

Figure 4.4. Influence of side pressure on rock compressive resistance

Figures 4.2 and 4.3 show charts of vertical stress measurement results and the change of the coefficient of average horizontal stress in relation to the average vertical stress, depending on the depth of occurrence.

The presented charts show that the horizontal components of stress state have various values. Carried in Great Britain research has shown that the horizontal components can differ depending on the direction (Doaws, Hons 1992). It has been shown that, depending on the localization, the maximum horizontal stress exceeds vertical stress from 20% to 100% (i.e. σ_z = 15 MPa; σ_x = 25 MPa; σ_y = 12 MPa).

In the light of the quoted research the vertical component of stress state is consistent, in terms of quality, with the classic approach, however the research results of horizontal stress is showing completely different ratios, depending also on the region of carried research. Generally it can be stated that the average horizontal stress relation to the average vertical stress on the H depth is in the range of (Hoek, Brown 1978):

$$\frac{100}{H} + 0{,}30 \le \frac{\sigma_x}{\sigma_z} \le \frac{1500}{H} + 0{,}50 \qquad (4.1)$$

The relation between vertical and horizontal stress has an essential significance to define rock limit state, and defining the parameters of supports cooperation with a rock strata. Studies have shown that compression of a sample rock in three-axis conditions (Kwaśniewski 1987; Kwaśniewski 2002), with the increase of side pressure, the rocks resistance increases (Fig. 4.4).

Analyzing the influence of depth on the size and distribution of stress, it can be stated that with the increase in depth in a certain region, significant changes in quality can occur. Having in mind that the influence of external factors on rock strata stress changes are expressed generally in relation to the stress value, the precision of their definition gains particular significance.

5. THE ANALYSIS OF HEADING LOCALIZATION DEPTH INFLUENCE ON ITS STABILITY

Generally used analytic solutions define the stress state in the neighborhood of headings with considerable idealization and simplification. Limitations and considerable idealization of actual conditions decided the choice for the definition of numerical modeling heading stability, with the usage of the finite element method. Limit state in the rock strata of the modeled disk is defined according to the Coulomb-Mohr criteria.

A serious problem during designing a heading on new levels is the lack of a sufficient reconnaissance of the massive. In this situation, to estimate the size and distribution of stress and stability maintenance conditions on a new level, data from active levels were used, and the results were approximated in function of depth.

Figure 5.1. shows the forming of a headings, which was made and maintained in individual layers in function of depth, support load capacity, and Figure 5.2. shows a chart of heading outline shift increase coefficient, depending on the depth of occurrence (the relation between the analyzed parameter on a determined depth and the value of this parameter on the depth of the existing level 700).

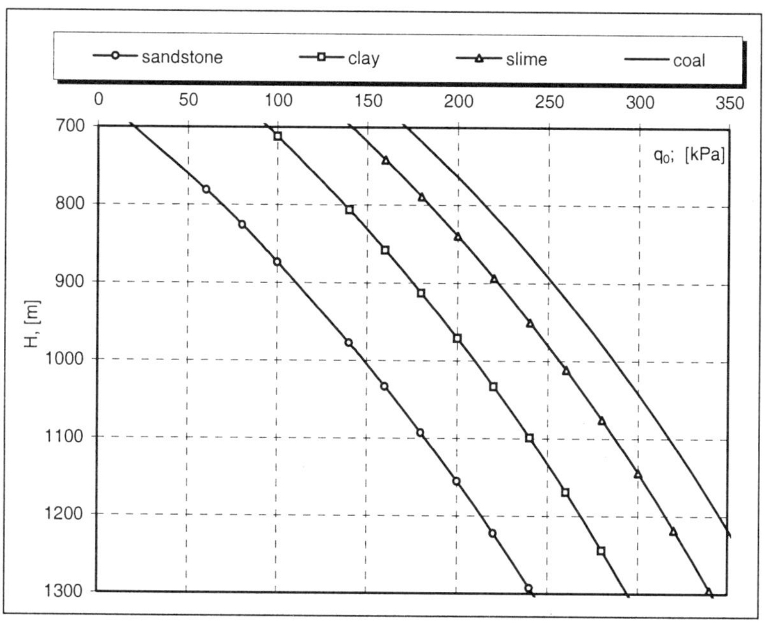

Figure 5.1. Chart of heading support load depending on depth (Duży 2007)

Referring the forecast size of stress on 1300 m level to the stress on 700 m level, an increase in stress in the direct neighborhood of the heading by approximately 30–50% can be expected. It can be a significant increase in rock, which build the massive, limit state and generate danger. The increase in rock strata limit state can cause the process of rock destruction and the occurrence of increased support load (even many times grater then on a higher level).

A considerable increase, with the increase in depth, is observed in the heading outline shift size. On 1300 m level, in relation to 700 m level, an increase in roof shift of approximately 70%, floor of approximately 80% and wall of approximately 100% can be expected. It points to a possibility of considerable intensification of the heading clamping process or other prophylactic action.

Having in mind the changeability of rock massive properties, the size and distribution of stress in the headings place of localization, calculations of headings depth of occurrence influence on heading stability were carried. To the calculations adopted geotechnic changeability parameters, based on the carried analysis, the support load capacity and its variance was adopted based on the statistical model presented in (Duży 2007). The evaluation of heading stability was carried for ŁP-10/A/ /V29 and ŁP-10/A/V36 support in a distance between the door frame of 1m. As the measure of reliability, the probability of stability loss, defined using the level II probabilistic method, assuming the range of heading localization depth in the boundaries of 700–1300 m, was adopted (Fig. 5.3).

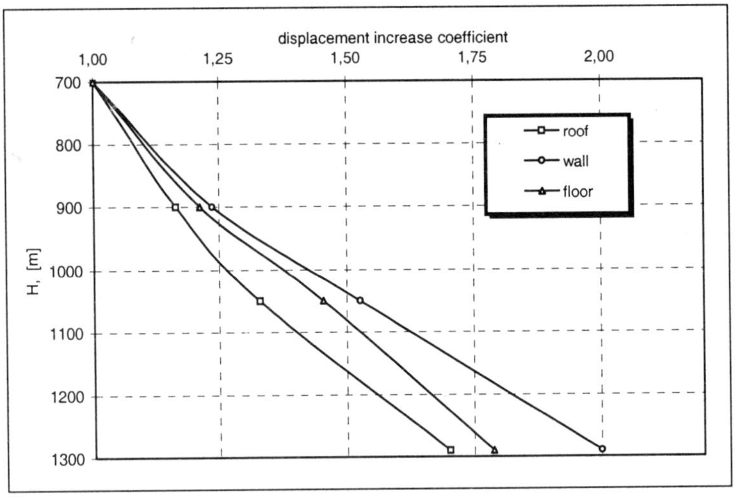

Figure 5.2. The forming of heading contour displacement increase coefficient depending on deposition of rock strata depth (Duży S. 2007)

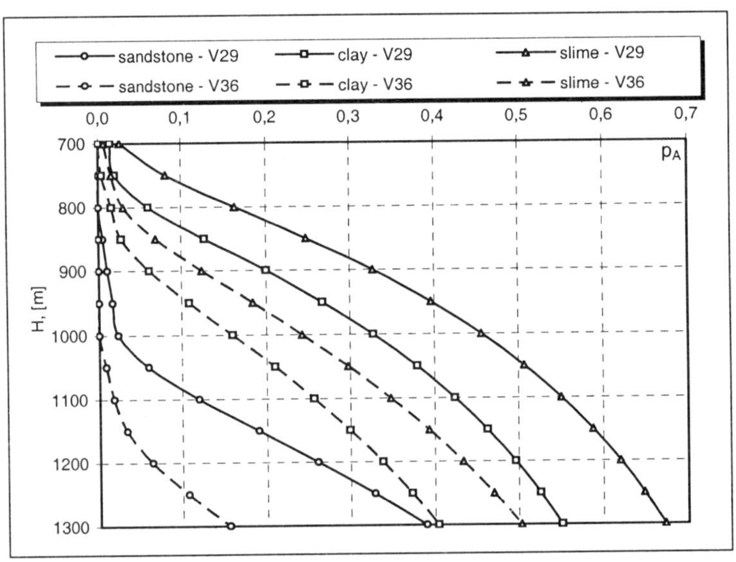

Figure 5.3. Heading support stability loss probability distribution depending on localisation depth (Duży 2007)

CONCLUSIONS

The basic factor that decides about the heading stability is an appropriately designed support, and one of its design main input parameter is its support load, which depends on the character of stress-sdeformation processes, which occur in the neighborhood of the heading. One of the most important factors deciding about the heading maintenance conditions are: the type and properties of the rocks, which build the massive, heading localization, the size of the cross-section, the type of the support, driving technology etc.

As the so far carried research showed, with the increase of depth, the danger of heading stability loss also increases. The increase in danger should be connected not only with the values of geostatic stress, but also with the increasing changeability of geological and mining conditions.

The presented research results showed that with the increase in designed and maintained headings localization depth, there will be more difficult maintenance conditions. On the increase of stability loss probability, decisive influences have:

- increasing, with the increase in depth, variance of rock strata basic geotechnic parameters values;
- increasing, with the increase in depth, variance of rock strata stress values in the surroundings of headings;
- increasing, with the increase in depth, probability of natural dangers occurrence;
- lesser precision of heading implementation, caused by the increased rock strata limit state.

Designing exploitation on greater and greater depths, signaled problems must be solved, so a parallel research of stress-deformation phenomenons changeability, in the neighborhood of headings, reconnaissance and a detailed description of research course decisive factors seems necessary.

REFERENCES

Brown E.T., Hoek E. 1978: Trends in Relationships between Measured In Situ Stresses and Depth. Int. J. Rock Mech. Min. Sci. & Geomech., Abstract, Vol. 15, Pergamon Press Ltd, 211–215.

Chudek M. 2002: Geomechanika z podstawami ochrony środowiska górniczego i powierzchni terenu. Wyd. Pol. Śl., Gliwice.

Chudek M. 1986: Obudowa wyrobisk górniczych. Cz. I. Obudowa wyrobisk korytarzowych i komorowych. Wyd. Śląsk, Katowice.

Dows G., Hons B.Sc. 1992: Kotwienie stropu w górnictwie węglowym – projektowanie i realizacja. Wiadomości Górnicze, nr 1, 27–32.

Duży S. 2006: Ocena zmienności własności wytrzymałościowych skał dla potrzeb projektowania budowli podziemnych. [W] Praca zb. pod red. M. Cały i D. Flisiak: Geotechnika i budownictwo specjalne. Wyd. AGH, Kraków, 37–48.

Duży S. 2007: Studium niezawodności konstrukcji obudowy i stateczności wyrobisk korytarzowych w kopalniach węgla kamiennego z uwzględnieniem niepewności informacji. ZN Pol. Śl., s. Górnictwo, z. 277, Gliwice.

Duży S. 2007: Wpływ głębokości lokalizacji wyrobisk górniczych na niezawodność i bezpieczeństwo ich konstrukcji. Bezpieczeństwo Pracy i Ochrona Środowiska w Górnictwie, nr 6 (154), 42–43.

Duży S. 2007: Wpływ głębokości lokalizacji wyrobisk górniczych na niezawodność i bezpieczeństwo ich konstrukcji. XI Warsztaty Górnicze pt. "Zagrożenia naturalne w górnictwie", Wyd. IGSMiE PAN, Kraków, 183–196.

Duży S., Szafulera K. 2006: Przestrzenna losowa zmienność wytrzymałości na ściskanie łupków piaszczystych towarzyszących pokładom węgla kamiennego warstw orzeskich. Górnictwo i Geologia, nr 2, 5–16.

Duży S., Szafulera K. 2007: Przestrzenna losowa zmienność wytrzymałości na ściskanie piaskowców towarzyszących pokładom węgla kamiennego warstw orzeskich. Budownictwo Górnicze i Tunelowe, nr 1 (w druku).

Duży S., Szafulera K. 2006: Przestrzenna losowa zmienność wytrzymałości na ściskanie łupków ilastych towarzyszących pokładom węgla kamiennego warstw orzeskich. XII Międzynarodowe Sympozjum "Geotechnika 2006", Gliwice, 171–181.

Filcek H., Walaszczyk J., Tajduś A. 1994: Metody komputerowe w geomechanice górniczej. Śląskie Wydawnictwo Techniczne, Katowice.
Harget G. 1973: Variation of Rock Stress with Depth at a Canadian Iron Mine. Int. J. Rock Mech. Min. Sci. 10, 37–51.
Hast N. 1969: The State of Stress in the Upper Part of the Earth's Crust. Tectonophysics, 8, 169–211.
Kropotkin K.N. 1972: The State of Stress in the Earth's Crust as Based on Measurements in Mines and on Geophysical Data, Phys. Earth Planet, Interiors 6, 214–218.
Kwaśniewski M. 1987: Badania nad mechanicznymi własnościami skał karbońskich dla potrzeb projektowania wyrobisk górniczych oraz prognozowania deformacyjnych i dynamicznych przejawów ciśnienia górotworu. Problem Resortowy MGiE, nr 119: "Metody i środki eksploatacji na dużych głębokościach", Wyd. Pol. Śl., Gliwice.
Kwaśniewski M. 2002: Zachowanie się skał izo- i anizotropowych w warunkach trójosiowego ściskania. ZN Pol. Śl., s. Górnictwo, z. 247, Gliwice.
Murzewski J. 1989: Niezawodność konstrukcji inżynierskich. Wyd. Arkady, Warszawa.
Polska Norma PN-ISO 2394:2000: Ogólne zasady niezawodności konstrukcji budowlanych.

International Mining Forum 2009, Sobczyk, Kicki & Saługa (eds) © 2009 Taylor & Francis Group, London, UK, ISBN 978-0-415-80428-8

Development Possibilities of Underground Coal Exploitation in Republic of Serbia

Zlatko Dragosavljević, Miodrag Denić
JP za PEU, Resavica, Republic of Serbia

Duško Đukanović
JP za PEU, Biro za projektovanje, Beograd, Republic of Serbia
e-mail: ugaljprojekt@sezampro.rs

ABSTRACT: Development perspective of Serbia is based on plans and development strategy of energetic mining, especially coal exploitation, which takes part in the production of electric power, over 66%.

Balance coal reserves in underground exploitation in Serbia, is about 280,000,000 t of coal in active mines, whereas according to the assessment there are about 235,000,000 t of potential reserves.

If you also take into consideration the promising deposits, which are examined partially, and still are not open, it is estimated that with underground exploitation, about 1,000,000,000 more t of coal can be excavated.

Accepting coal reserves and development strategy of energetic sector, it is necessary to plan the construction of new steam power plants, which would use coal from underground exploitation.

The development possibilities of underground coal exploitation in Serbia are given in this paper.

KEYWORDS: Underground coal mining, reserves, coal

INTRODUCTION

Serbia development perspective is based on plans and strategy of energetic mining development, especially coal exploitation, which is over 66%, present in electric power production.

Coal production is done in open pit mines, underground exploitation mines, and in sub aqueous exploitation mines. Over 95% of production's purpose is for the needs of burning in steam power plants and for the transformation into electric power. The needs of Serbia for the coal for the mass consumption are satisfied in amounts of about 6 million tones a year. In Table 1 the placement of coal is shown according to the source of supply in Serbia.

Table 1. Coal placement according to source of supply

Coal placement according to supply source	10^6 t	participation, %
Surface exploitation mines	36	93,8
Underground exploitation mines	0,65	1,69
Sub aqueous coal exploitation	0,2	0,52
Coal import	1,53	3,98
TOTAL	38,38	100

The coal reserves in underground exploitation of Serbia in active mines are about 515.000.000 t of coal. If you take into consideration perspective deposits which are researched partially, and not open yet, the assessment is that with the underground exploitation about 1.000.000.000 t of coal can be excavated. 2,5 billions of tones of oil slates of Aleksinac basin can be added to this, out of which 200.000.000 t of oil can be obtained.

Accepting the coal reserves and strategy of energetic sector development, the development possibilities of underground coal exploitation in Serbia.

RAW MATERIAL BASE – COAL RESERVES

Within this point the raw material potential in active coal deposits which are in a process of exploitation will be given, potential deposits in which there was no exploitation executed and raw material deposit base in which formerly the exploitation works were stopped.Stone coal deposits ("Vrška Čuka" and "Ibar mines") belong to geological formations of Trias, that is Miocene. Deposits "Rembas", "Soko", "Bogovina" and "Jasenovac" belong to geological formations of a lower and middle Miocene. Deposits "Stavalj" and "Lubnica" are of Miocene Age, whereas the coal from the deposit "Stavalj" is by its characteristics closer to dark coal than to brown coal. Coal in deposits apart from exploitation belongs to different geological formations and those are: Rtanj – stone coal, Aleksinac and Nova Manasija – brown coal, while the coal from the deposit Poljana and Melnica belongs to brown coals.

According to coal reserves state on day 31.12.2006. of total balance coal reserve in deposits in exploitation is 279.275.400 t, and out-of-balance coal reserves in deposits in exploitation is 16.803.680 t as shown in Table 2.

Table 2

Coal/Mine	Balance reserves, t $A+B+C_1$	Out-of-balance reserves, t $A+B+C_1$	Potential reserves, t C_2
Stone coal			
Vrška Čuka	2.361.230	350.000	6.000.000
Ibarski rudnici	2.632.540	1.223.980	1.000.000
Σ Stone coal	4.993.770	1.573.980	7.000.000
Dark coal			
Rembas	12.207.330	540.060	15.000.000
Bogovina	2.058.260	1.897.190	1.000.000
Soko	58.127.960	2.763.270	150.000.000
Jasenovac	1.210.900	0	8.000.000
Štavalj	187.086.180	7.709.550	50.000.000
Σ Dark coal	260.690.630	12.910.070	224.000.000
Dark-brown			
Lubnica	13.591.190	2.319.630	4.000.000
Σ Dark-brown	13.591.190	2.319.630	4.000.000
Σ JP PEU	279.275.400	16.803.680	235.000.000

Total balance coal reserves in deposits apart from exploitation are about 169.813.362 tones, out-of -balance 21.539.107 tones and 57.000.000 tones of potential reserves, as shown in Table 3.

Table 3

Coal/Mine	Balance reserves, t A+B+C_1	Out-of-balance reserves, t A+B+C_1	Potential reserves, t C_2	Geological reserves, t
Stone coal				
Rtanj	1.579.440	39.060	1.000.000	2.618.500
Bajovac	26.530.962	1.502.137	0	28.033.099
Σ Stone coal	28.110.402	1.541.197	1.000.000	30.651.599
Dark coal				
Aleksinac	12.320.190	15.195.430	10.000.000	37.515.620
Nova Manasija	3.351.000	934.000	20.000.000	24.285.000
Σ Dark coal	15.671.190	16.129.430	30.000.000	61.800.620
Brown coal				
Despotovac	27.956.970	684.480	-	28.641.450
Melnica	39.537.400	-	10.000.000	49.537.400
Poljana	58.537.400	3.184.000	16.000.000	77.721.400
Ćirikovac	-	-	-	200.000.000
Σ Brown coal	126.031.770	3.868.480	26.000.000	355.900.250
Σ Potential	169.813.362	21.539.107	57.000.000	448.352.469

Total geological reserves of deposit Ćirikovac are given in the table.

Potential reserves in active deposits are estimated to 235.000.000 t, adding to this and coal reserves in deposits in which there was no exploitation executed, as well as the deposits in which the exploitation is stopped temporarily, it is estimated that the total raw material coal potential for underground exploitation is about 1.000.000.000 t.

UNDERGROUND COAL EXPLOITATION PERSPECTIVES

With underground coal exploitation, high-calorie coals are exploited, which are mainly used for energetic purposes and a small part as a technological raw material:

- semi-anthracite, is used as a technological raw material for the active coal production, and then as a filler and a raw material for the coax and semi-coax . It is used in casting industry, in cement production and brick-making industry and industrial boiling stations;
- stone coal is used mainly in industrial boiling stations, steam power plants and casting and brick-making industry;
- dark coal and dark-brown coal is used in mass consumption (households and block-boiling stations), in industrial boling stations, heating plants, steam power plants, cement industry and brick-making industry.

Taking into account the current state and JPPEU organization, it is necessary to have a new approach.

In the fields of technique and technology of exploitation the number of measures is to predict by which the production and productivity are to be increased, safety measures to be improved and a hard miner work to be humanized. In basis the modernization of technological processes is necessary by introduction of new equipment and procedures in particular phases of exploitation, which is the condition for underground coal exploitation.

The measures towards the optimization of the basic elements of technique-technological system are to be executed in mines, such as:

- Intensifying of production works in shafts, pits and mines;
- Modernization of technological procedures of underground halls construction of opening, working out and preparation;
- Intensifying of geological research works, from surface and from pits where the greatest volume of works should be realised from one's own operative;
- To mechanize the work phases and adjust them to concrete natural-geological deposit conditions;
- Rationalization of the excavation transportation system, people transportation, production material and equipment delivery;
- Modernization of processing procedures and valorisation of small assortment of coal;
- Improving of maintenance system, servicing, production material preparation, technical protection system, ventilation and drainage, as well as other measures of work humanization;
- Improving of work organization and management system improvement, labour reduction to the optimally sufficient number, without enormous administrative service and many irrelevant procedures;
- Education of the qualified structure of employees and improvement of specialized knowledge of the management personnel.

Accepting the geological reserves and territorial position of mine "Lubnica" and mine "Soko" and Aleksinac deposit with the known coal quality possessed by the mines, the construction of the steam power-heating plant is suggested, of power 200 MW. For the realization of that goal it is necessary to introduce the mechanized excavation in "Lubnica" mine, in order to achieve the production of 200.000 tones of coal annually. In "Soko" mine it is necessary to introduce the mechanized excavation with short long-walls for the reason of the complex geological conditions, so to achieve the production of 300.000 tones of coal annually. For the realization of the above mentioned it is necessary to do the underground halls construction in "Lubnica" mine about 1000 m, and in "Soko" mine about 2500 m, for the realization of the planned investments for the achieving the mentioned production capacity the investments of 5.000.000 euros for the "Lubnica" mine are necessary, and for the "Soko" mine 20.000.000 euros. In Aleksinac deposit, production rehabilitation is necessary, on the level before the mine closing.

Accepting the geological reserves and territorial position of the "Rembas" mine and "Jasnovac" mine, with the known quality of the coal that is possessed by the mines, it is suggested that the coal be placed in steam power plant Morava-Svilajnac, which is revitalized until recently, and thus its work life prolonged. For the realization of that goal it is necessary to introduce in "Rembas" mine the mechanized excavation with the long-wall in "Strmosten" pit, in order to achieve the 300.000 tones production of coal annually for "Rembas". In "Jasenovac" mine it is necessary to maintain the production for 70.000 tones of coal annually. For the realization of the mentioned it is necessary to do the underground halls construction in "Rembas" mine, pit "Strmosten" about 3000 m. For the realization of the planned investments for the achieving of the mentioned production capacity 15.000.000 euro investments for the "Rembas" mine are needed.

Accepting the geological reserves and territorial position of the "Štavalj" mine, with the known quality of coal possessed by the mine, the construction of steam power plant in Sjenica is suggested, of power 300 MW. For the realization of the mentioned it is necessary to increase the production (introduction of the mechanized excavation) and opening of new production capacities. For the realization of the mentioned 10.000.000 euro investments are needed.

Coal production from Ibar mines can be placed in the future TE-TO Sjenica, or in Energetics Kragujevac or in TE Morava. In Ibar mines it is necessary to do the conservation of "Jarando" pit in cooperation with the public institutions, and as a replacement capacity for opening the open pit "Progorelica".

For the mines Bogovina and Vrška Čuka, it is necessary in cooperation with the public institutions to transfer to conservation of pits phase.

The mentioned approach would enable reaching the production capacity from JPPEU mines on level from about 1,5 mil. tones annually.

Apart from the present active deposit coal mines, there are other numerous coal deposits in Serbia that possess the significant researched coal reserves which can be activated and give contribution to problem solving in fuel supplying and mass consumption. Deposits which are not exploited represent the significant raw material energetic base, where the increased production capacity can be developed, which shows the need for opening of new mines especially in deposits with greater reserves and suitable natural-geological conditions as according the general deposit assessment: Poljana, Melnica, Despotovac basin, Kosa-Zabela, Zapadno-moravski basin, deeper parts in great brown-coal basins (Kolubara and Kostolac-Ćirikovac) and numerous smaller deposits. Here also, the reserves of oil slates of Aleksinac region should be taken into account, which is estimated at about 2.500.000.000 t.

CONCLUSION

Accepting the balance coal reserves, the current state of energetic sector in Serbia, the perspective of the mine with the underground coal exploitation is reflected in the construction of new thermo-energetic capacities. The construction of the new thermo-energetic capacities imposes the modernization and rationalization of production in coal mines, as well as the openings of new mines with the underground coal exploitation.

REFERENCES

[1] Ristović I.: Restrukturiranje i privatizacija rudnika uglja sa podzemnom eksploatacijom u Srbiji. Zbornik radova I Međunarodni simpozijum Energetsko rudarstvo 2007, str. 15–21, Vrnjačka banja 2007.
[2] Strategija energetskog razvoja republike Srbije do 2015. godine.
[3] Nacionalna strategija privrednog razvoja do 2012. godine.
[4] Projektna dokumentacija JPPEU-Biro za projektovanje i razvoj. Beograd.

International Mining Forum 2009, Sobczyk, Kicki & Saługa (eds) © 2009 Taylor & Francis Group, London, UK, ISBN 978-0-415-80428-8

Actual State of Underground Coal Exploitation in Republic of Serbia

Zlatko Dragosavljević
JP za PEU, Resavica, Republic of Serbia

Duško Đukanović
JP za PEU, Biro za projektovanje, Beograd, Republic of Serbia
e-mail: ugaljprojekt@sezampro. rs

ABSTRACT: Mines with underground coal exploitation in Republic of Serbia operate within the Public enterprise for underground coal exploitation Resavica, which is owned by Republic of Serbia 100%.

Serbian coal mines with underground mining represent low-volume manufacturing plants. The overall production of coal mines with underground exploitation is carried out in 8 mines with 11 pits. The quality of coal is among high rated and such coal is used for consumer's and industrial consumption.

This paper shows the actual state of underground coal mining in Serbia.

KEYWORDS: Underground coal mining, coal

INTRODUCTION

Underground exploitation mines in Republic of Serbia act within the Public enterprise for underground coal exploitation Resavica, the overall coal production is carried out in 8 mines with 11 pits. The coal falls within those types of the highest quality and is used for a wide and industrial consumption.

Coal exploitation is carried out with pillar methods which have a low productivity. The currently available equipment of the Mines is very amortized and not reliable, which affects the reduction of work efficiency additionally and which is very low.

Former practice of keeping the technological process of underground coal exploitation on a low degree of mechanization with the high participation of physical work, lack of investments, price policy (many-years inconvenient coal price) and material discouragement of the underground mining work brought the coal mines with the underground exploitation to a bad position.

The current state of affairs in underground coal exploitation in Serbia is given in this paper.

GEOGRAFICAL POSITION OF THE UNDERGROUND COAL EXPLOITATION MINES

Underground coal exploitation mines in Serbia, do business within the Public Enterprise for the underground coal exploitation in Serbia (JP PEU) with the headquarters in Resavica, which is 100%

owned by the Republic of Serbia. JP PEU is a unique Publics Enterprise and the production is carries out in 8 mines with 11 pits in: hard coal mine Vrška Čuka, Avramica, Ibar mines of the stone coal – Baljevac, Rudnik mrkog uglja Rembas – Resavica, dark coal mine Bogovina – Bogovina, dark coal mine Soko – Soko banja, dark coal mine Jasenovac – Krepoljin, brown coal mine Lubnica – Lubnica, brown coal mine Štavalj – Štavalj, and Aleksinac mine (mining investment works).

Hard coal mine Vrška Čuka is situated in Eastern Serbia at about 10 km from Zaječar, alongside the Bulgarian border. Ibar mines of the stone coal are situated in the Ibar valley, southward from Kraljevo. Within the Ibar mines pits Jaran Do and Progorelica work and open-pit mining Tadenje.

In Serbia there are four active mines of dark coal, Rembas, Bogovina, Soko and Jasenovac.

Senjsko-resavski coal basin is situated in western parts of Eastern Serbia and spreads from village Panjevac in north to Čestobrodica in south. It is situated at about 23 km distance from the highway Niš–Beograd in direction of Ćuprija or 18 km from Despotovac, that is 40 km from Svilajnac. In Rembas mine there are 3 active pits: Strmosten, Senjski Rudnik and Jelovac.

Bogovina coal basin is situated in Eastern Serbia, on foothill of Južni Kučaj. Dark coal mine Bogovina is situated at about 10 km from Boljevac,and at about 25 km of air line from Bor. Dark coal mine Soko is situated at about 12 km from town Soko Banja, or at about 30 km from Aleksinac and highway Niš–Beograd and represents a part of Sokobanja coal basin.

Dark coal mine Jasenovac belongs to Krepoljin basin and is situated in Krepoljin–Sige–Krupaja area. Jasenovac mine is situated at about 62 km from Požarevac.

There are two dark-brown coal active mines with underground exploitation in Serbia.

A dark-brown coal mine deposit Štavalj represents a part of Sjenica coal basin. It is situated at about 12 km from Sjenica that is at about 85 km from Novi Pazar.

Coal area Lubnica belongs to Lubnica-Zvezdans coal basin which spreads between Beli and Crni Timok, not far away from Zaječar, through Zvezdan and Lubnica to Osojna. Mine Lubnica is situated at about 8 km from Zaječar.

JP PEU is an enterprise which is located in 9 municipallities (Zaječar, Raška, Kraljevo, Despotovac, Soko Banja, Boljevac, Žagubica, Sjenica, Aleksinac). In Figure 1 the locations of mines which work within JP PEU are shown.

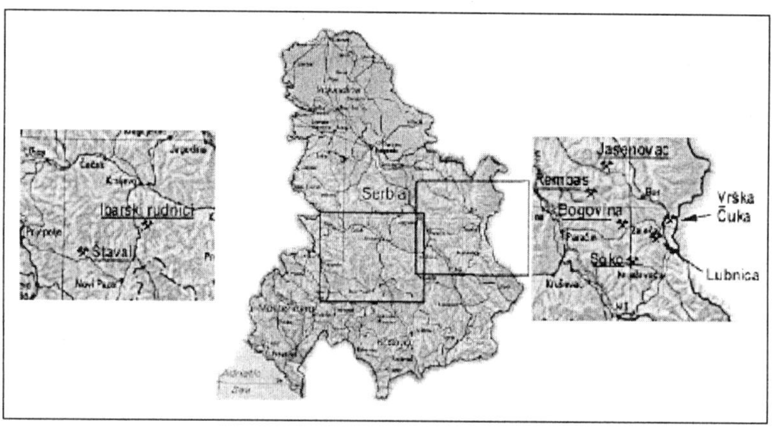

Figure 1. Locations of mines which work within JP PEU

Such displacement of production plants affects and significantly increases the price of executing the basic business functions of management and leadership, production-technical functions, economic-financial, personnel, legal and other functions and businesses of significance for the effective functioning of JP PEU.

ORGANIZATION STRUCTURE OF JP PEU

JP PEU in today's shape functions from 2003. JP PEU was an integral part of JP EPS until 2003. On basis of the decision of the Serbian Government from 15.05.2003, JP PEU is separated from EPS and acts as JP PEU-Resavica. Coal mines as its part are organized as the enterprise parts, and the enterprise management with the headquarters in Resavica is organized in a way so to unite the basic business functions and creates conditions for unabrupt executing of works on coal exploitation. At that moment there were about 5.500 employees whereas today there are about 4.200 employees working in it.

Organization scheme JP PEU is shown in Figure 2.

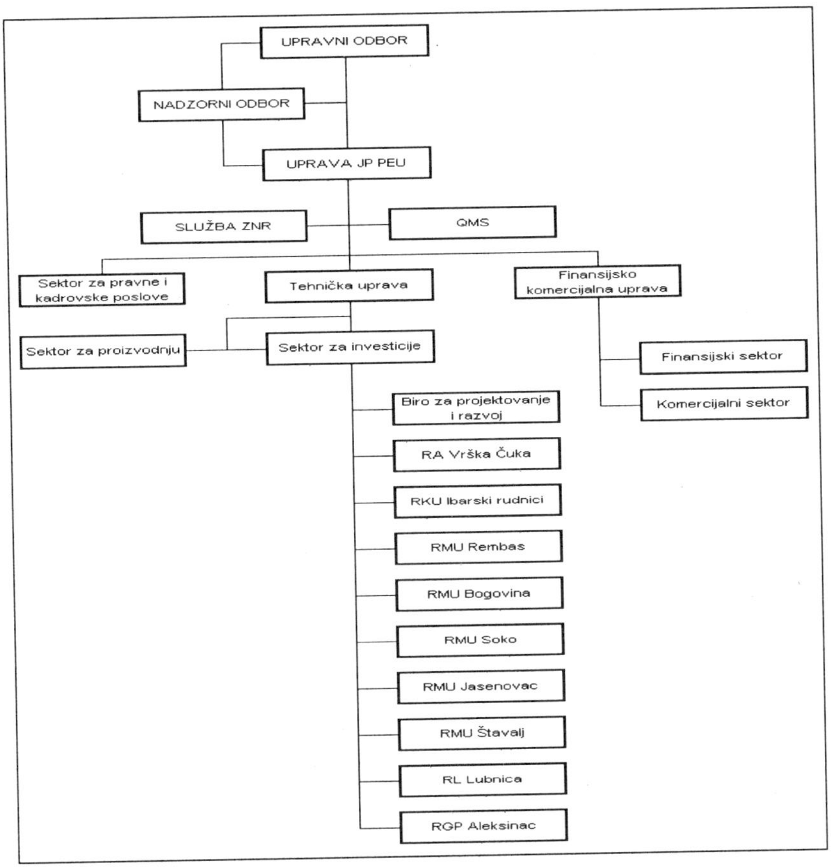

Figure 2. Organization scheme JP PEU

NATURAL-GEOLOGICAL CONDITIONS IN JP PEU MINES

Natural-geological conditions which determine the conditions of the exploitation in active coal deposits are characterized by the following:

- Tectonic conditions in all deposits are complex with the dominant tectonic deformations whose consequences are deformed shapes of the limited exploitation areas with the relatively short lengths of excavation fields and frequent changes of the direction of spreading and angles of seam slopes.

- According to depth of dip and strike of coal seams, most of the deposits belong to the mine group with the average exploitation depth (to 400 m).
- Hydrogeological conditions are changeable and it is about the small inflows of water into underground objects, where the exception is the deposit of the "Štavalj" mine with the bigger inflow of water (over 3 m^3/min).
- Active deposits are not necessarily the bearers of methane. Greater amounts of methane are in "Soko", "Vrška Čuka" and "Jarando" mines.
- In floor and roof of coal seams dominate the rocks with prevailing participation of clay components, with low values of mechanical characteristics, which provokes rock boosting and deformation of the support of the constructed underground mining halls.
- The quality of coal is in a wide range of values with the ecologically acceptable contents of hazardous components and warmth values adequate for use in industry and wide consumption. Coal seams which are the subject of exploitation in active pits, except for the "Vrška Čuka" deposit, are prone to self-inflammation, and coal dust shows hazardous characteristics.

Present natural-geological conditions and lack of investing and mechanizing had an impact on the choice of technological solutions of exploitation, so that today in all active pits classic pillar methods of excavation are used in different variants, where the working phases are half mechanicized and production effects different, mainly during the different exploitation conditions.

The degree of a research and degree of the knowledge of the working environment conditions is on a very low level, which makes designing and leading of technological production process hard.

The available equipment in mines, except from a few exceptions, is outdated, which is the consequence of the objective circumstances (former sanctions, price policy and current lack of means etc.) but as well as subjective ones (neglecting of transportation problem, lack of concept etc.),

Hard working conditions caused by low mechanization and material discouragement causes a permanent reduction of a pit labour. The interest of a new labour for the employment in mines was almost symbolic, until last year when the interest for the employment in mines grew.

Production level reduction and a low productivity is mainly connected to the ways of underground halls construction, excavation method and transportation system. Underground halls construction is carried out with classic drilling-blasting method. In technological excavation process not keeping up the pace with the use of new technologies is more obvious. In last ten years there was no work with mechanized excavation in pits and apart from the existing conditions for their use in deposits that is in parts of deposits. Particularly there is a problem with the transportation system and production material delivery and people transportation, which because of aging causes the great number of workers for these jobs, and at the expense of other technological phases.

The state of a personal and collective protection in mines with underground coal exploitation is followed by significant difficulties, which are mainly connected to the lack of foreign financial means for import of the necessary equipment and instruments. Mainly in all pits with the methane regime of work there are installed systems for automatic gas-ventilation parameters control and voice-speaking communication, whose functionality is connected to a regular maintenance.

FORMER PRODUCTION OVERVIEW

Underground coal exploitation in active deposits is characterized by continual fall of production level in last decades, which were influenced apart from the complex geological conditions, by the very low degree of mechanization of working phases of a technological process, outdated fixed assets, insufficient researching of the working environment conditions, unsuitable structure of labour and insufficient investments in rehabilitation and development of phe production capacities.

Table 1 data show the production of coal per five-year periods (from 1965 to 1990) and per years (from 1991) in present active mines.

Table 1. Achieved production overview-coal mines with underground exploitation

Year	Production, t	Year	Production, t
1965	1.305.300	1970	1.476.096
1975	1.300.269	1980	1.035.924
1985	1.035.870	1990	861.571
1991	954.180	2000	623.230
1992	960.973	2001	549.159
1993	687.765	2002	540.741
1994	686.393	2003	540.658
1995	662.445	2004	534.261
1996	651.239	2005	551.960
1997	697.943	2006	491.809
1998	634.013	2007	450.800
1999	579.202	2008	513.000

It is obvious from the Table 1 data that there is a significant fall of coal production from underground exploitation, which is the consequence of more causes, and the most significant are:

- the fall of coal production level from underground exploitation starts from 60's when at the expense at that time "cheap oil", there was an elemental closing of mines with underground exploitation and forced oil consumption and oil derivatives in industry;
- coal replacement as a fuel in industry, heating plants, wide consumption and traffic first of all liquid and gas fuels and electric power;
- wars in the region and sanctions against Serbia. This period is characterized by the start of a more significant fall of production in underground coal exploitation mines. Hard production conditions which were caused by the events from that period, had a great impact on methods of excavation, equipment, maintenance and more other factors significant for the work effectiveness and business effectiveness to stay on a former development level, outdated and ineffective;
- separating the JP PEU from JP EPS and the price policy.

Separating JP PEU from JP EPS resulted in foundation of a new enterprise with the accumulated loss of 4.848.826.000 dinars, taking into consideration that the new enterprise should be formed without loss and liabilities, so burdened it took over the costs of restriction and current production which deteriorated current liquidity more.

Table 2

Year	Amount, t
2003	110.855
2004	170.477
2005	154.406
2006	118.394
2007	106.817
2008	150.000

Present capacity of the active mines in normal circumstances with the existing shafts is 600–700.000 t/year, under condition of providing the financial assets for the equipment purchasing and production material to support the current production and the parallel investments in few excavation shafts.

The main coal consumers of the mines with the underground exploitation are steam power plants, industry, heating plants, households and public institutions. Great amounts of coal are sold in commercial enterprises, which supply households, industrial and other consumers in local areas. As for the coal selling to the power plants the main buyer is stem power plant Morava. Table 2 shows the amounts of coal which are delivered to the steam power plant Morava in period from 2003 to 2008. For 2009 a delivery of 260.000 t of coal is planned.

Coal market of Republic of Serbia is deficient in types of coal with the high quality of coal because their consumer basis goes up to two million tones annually. Even in today's, atypical times, when the industrial and economy activity is significantly reduced there is a significant lack of a good quality coal on the market.

BASIC PRODUCTION PARAMETERS OVERVIEW FOR THE PERIOD 2001–2008

Coal mines business is overburdened with chronically lack of the current assets for normal current business, lack of means for the development and providing the new excavation capacities, unsolved issues of subsidiary and service business burdened with a considerable number of unproductive labour, as well as inadequate coal price which don't follow the price growth input. Coal prices in the previous period are under control, so that it considerably influenced the economic position of the mine. At the same time, production material and equipment prices were enlarged greatly and followed the market currency. Significant reduction of subvention of coal production with underground exploitation, it influenced the total business of mines with underground exploitation in great measure. In Table 3 and Figure 3 the parameters of coal production from underground exploitation are given for the period from 2001 to 2008.

Table 3. Coal production parameters from underground exploitation from 2001 to 2008

Year	2001	2002	2003	2004	2005	2006	2007	2008
Production, t	549.159	540.741	540.658	534.260	551.960	491.809	450.800	513.000
Number of employees	5.500	5.413	5.248	5.291	4.481	4.280	4.230	4.150
Productivity, t	99,8	99,9	103,0	101,0	123,2	114,9	106,5	123,61

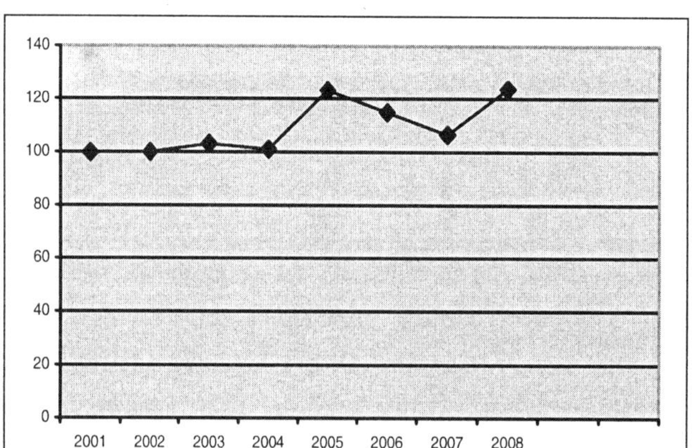

Figure 3. Productivity diagram t/employee for the period from 2001 to 2008

Accepting the productivity growth in 2008 the production of 570.000 t is planned for 2009 in underground way (in case of opening of two excavation sites "Progorelica" and "Biljkina Struga" the production of 660.000 tis planed) with the parallel reduction of labour to 4.000 employees. Planned productivity for 2009 is 142,5 t/employee (165 t/employee).

In period from 2001 to 2008 JP PEU has always had expenditures of the investment character, more precisely because of investing for the maintaining of current production. If it were not for the investment expenditures, or if there were other sources for the investment, JP PEU would have a need only for the liquidity subventions. Investments of priority provide the follow-up of business and production volume achievement.

The coal price in the former period was under control, so that it has had a significant impact on the mines economic position. Last increase of coal price (large assortments)was in 2006, while the last increase of coal price (small assortments) was in 2004.Current sales price of coal from underground exploitation is 1,7 eur/GJ. Simultaneously the prices of a production material and equipment are significantly enlarged and they followed the market currency. Table 4 gives the relation of prices of the basic production material (steel support, oil, explosive, electric detonators) for 2004, 2006 and 2008.

Table 4. Price relation of the basic production material for the years 2004, 2006 and 2008

Name	price 2004, RSD	price 2006, RSD	price 2008, RSD	increase, %
1	2	3	4	4/2
Steel support	45.245,96	52.690,21	76.888,43	69,93
Oil	50,9	67,95	94,1	84,87
Explosive	69	78,25	109,8	59,13
Electric detonators	46,36	49,84	53,52	15,44
Electric power	1,43	1,75	3,416	238,88

As for the comparison, meanwhile there were changes in coal price three times in the world, and we had the change in electric power price and not of coal.

Taking into consideration the permanent illiquidity and non-cost-efficiency of business of the mines with underground coal exploitation in Serbia, Ministry of Economy of the Republic of Serbia started the procedure of privatization of JP PEU, by the initiative from 13.03.2006. Privatization agency of Republic of Serbia made a decision on 24.03.2006 about initiating the procedure of reconstruction of JP PEU, and for the underground coal exploitation mines privatization in Serbia.

CONCLUSION

Current state in underground coal exploitation is conditioned by a range of subjective and objective circumstances. Coal price policy on the level of keeping the social peace, influenced directly the mine business economy (the mines purchased the necessary production material at market prices, and coal was sold at a lower price). Currently in JPPEU mines, the reduction of labour with the gradual increase of production and making the conditions for increasing the production and productivity considerably in the coming year, is obvious. It is crucial that the State has to change the price policy and use the underground coal exploitation as a strategic branch of economy and has to enable reorganization of the existing mines and a gradual opening of the new ones.

REFERENCES

Ristović I.: Restrukturiranje i privatizacija rudnika uglja sa podzemnom eksploatacijom u Srbiji. Zbornik Radova, I Međunarodni Simpozijum Energetsko Rudarstvo 2007, str. 15–21, Vrnjačka banja 2007.

Projektna dokumentacija JPPEU-Biro za projektovanje i razvoj. Beograd.

International Mining Forum 2009, Sobczyk, Kicki & Saługa (eds) © 2009 Taylor & Francis Group, London, UK, ISBN 978-0-415-80428-8

Modernization of Technological Process of the Construction of Shaft Premises in Coal Mines of Serbia

Duško Đukanović
JP za PEU, Biro za projektovanje, Beograd, Republic of Serbia
e-mail: ugaljprojekt@sezampro.rs

Miodrag Denić
JP za PEU, Resavica, Republic of Serbia

Dušan Dragojević
JP za PEU, Biro za projektovanje, Beograd, Republic of Serbia
e-mail: ugaljprojekt@sezampro.rs

ABSTRACT: The current state and analysis of applicable technologies for the construction of shaft premises in coal mines of Serbia is given in the paper. The proposal for modernization of technological process of the construction of shaft premises in coal mines of Serbia.

KEYWORDS: Shaft, mines, coal

INTRODUCTION

Shaft premises in Serbian underground coal mines are currently constructed with drilling and blasting technology, which is not satisfactory in matters of speed and cost of production. It is necessary to work on modernization of technological process of coal exploitation in mines of JPPEU in future period by introducing mechanized methods of excavation which will increase production of coal greatly, that is, the speed of exploitation field excavation. As result there is a need for faster and safer construction of shaft premises which has to follow the dynamics of excavation. Faster and safer construction of shafts premises can be achieved by introducing of modern technologies of constructing and supporting of shaft premises and that is, first of all, the application of combined machines for construction and rockbolts or some other modern support for supporting of premises.

The current state of affairs and analysis of applied technologies of shaft premises construction in coal mines of Serbia is given in this paper. There is a proposal for modernization of technological process of shaft premises construction in coal mines of Serbia.

CONSTRUCTION OF SHAFT PREMISES IN COAL MINES OF SERBIA – CURRENT SITUATION

Shaft premises in underground coal mines in Serbia are currently constructed in different working environments. The most common ones are: coal, waste rock (clay, sand-clay, clay-smarl, marl). Coal represents a good working environment for construction of shafts premises, whereas waste rock is a very ungrateful environment for the construction of shafts premises.

The most usual shapes of shaft premises cross section in Serbian mines are trapezoid (17%), domed (22%), arched (25%) and circular (36%). Currently, the shaft premises are constructed with trapezoid and circular cross section.

The most common types of supports for supporting of shaft premises are wood-trapezoid (11%), steel-circular (36%), steel-arched (22%), concrete (18%) and other types of support (13%). Currently, the shaft premises are supported with both steel and wood trapezoid and steel circular support.

The most common sizes of a cross section of shaft premises are: 7,5 and 8,6 m^2 (for trapezoid cross section), 9,62 and 12,56 m^2 (for circular cross section) and 8,6 and 10,75 m^2 (for arched cross section).

Constructed shaft premises are mostly horizontal and inclined with inclination lower than 16 in most cases.

ANALYSIS OF APPLIED TECHNOLOGIES OF SHAFT PREMISES CONSTRUCTION

JPPEU mines apply or applied technologies of constructing the shaft premises with drilling and blasting technology and a combined machine (roadheader). Shaft premises in Serbian underground coal mines are currently constructed with drilling and blasting technology.

In order to analyze the technologies of shaft premises construction with the drilling and blasting method, the recording of a technological process was executed in 4 shafts ("Lubnica", "Soko", "Strmosten" and "Jasenovac"), and after a statistic data processing the next parameters were given (Tab. 1).

Table 1

Time of work	418 (min/shift)	87,08%
Time of holdup	62 (min/shift)	12,92%
Construction speed	0,587 (m/shift)	
Efficiency	0,147 (m/wages)	

From Table 1 you can see that the achieved construction speed of shafts premises in coal environment is 0,587 m/shift, and that the achieved efficiency is 0,147 m/wages. Also you can see from Table 1 that the participation of holdups is about 13%, of total duration of the shift. Such a great percentage of holdups is connected to transportation system which is about 70% of total holdups. Reason for this is the amortized transportation equipment which is not maintained properly.

In coal mines in Serbia ("Rembas" – shaft "Jelovac", "Aleksinac", "Soko", "Štavalj"), roadheaders (F-6A i AM-50, PK III) were used formerly, which showed different results in its work. Within the paper there is an analysis of the results which are achieved during the construction of shaft premises with a roadheader type F-6A in shaft "Jelovac" (Tab. 2).

Table 2

Time of work	375 (min/shift)	78,13%
Time of holdup	105 (min/shift)	21,87%
Costruction speed	1,5 (m/shift)	
Efficiency	0,2 (m/wages)	

You can see from Table 2, that the accomplished shaft premises construction speed in coal working environment is 1,5 m/shift, and that the accomplished efficiency is 0,2 m/wages. You can also see from Table2 that the participation of holdups is about 22%, of duration of the shift. Such a great percentage of holdups is connected to the transportation system and production materials delivery as well as to the damages on the roadheader itself which participate with 15% of total holdups.

Such small amounts of the acomplished speed and efficiency of shaft premises construction in JPPEU mines is a result of inadequate work organization, inadequate and inconvenient equipment which is used in working conditions, insufficient training of workers in shaft premises construction field, lack of work conditions knowledge, inconvenience of technological process of construction and working operations versus mining-geological conditions etc.

The data which are given in Tables 1 and 2 are the most significant marker of the state of shaft premises construction in our coal mines. Apart from the reasons which influenced very low construction speed and efficiency, the fact is that the results which are achieved are far below those which represent the economic rationality and technical optimum.

SHAFT PREMISES CONSTRUCTION PERSPECTIVES IN COAL MINES OF SERBIA

Increase of shaft premises construction speed with the drilling and blasting technology method can be achievable by improvement of certain working operations. Improvement of working operations as drilling mine holes, charge and mine burning, can increase the speed of construction of shaft premises to 2,4 m/day with construction in working coal environment 1,5 m/day with construction in working waste rock environment.

The working environment where the shaft premises are constructed in coal mines of Serbia is suitable for the mechanized construction too, because the strength coefficient does not exceed 5, the mechanized construction can be used in working environment with the strength coefficient up to 12, too.

Constraining factors of justifiability for using the roadheaders for the shafts premises construction are the length of the premise which is constructed and the construction speed.

On basis of the criterion of the lowest direct costs of shaft premises construction by roadheaders, the minimum length of shaft premises in which roadheaders can be used are determined and they are: 400 m for shaft premises of trapezoid shape and 300 m for shafts premises of circular and arched shape. The speed of construction in a working coal environment which can be achieved by the use of the roadheader in the conditions of our coal mines is 6,4 m/day. The given data refer to the existing shapes, the dimension of a cross section and a type of the support. The speed of the shaft premises construction can be increased if some of the modern types of support material are used. Existing type of a support and the way of a support takes 1/4 of working cycle and slow down the construction process considerably.

Except that the existing way of supporting influences the construction speed, more frequent rehabilitations impose a question of justifiability of the use of former way of supporting.

Over the recent years in world a technology of supporting the underground premises by rockbolting with two-component quickharden mixture or shorter is used, AT (Advanced Technology) rockbolting, which not only replaces successfully the traditional way of supporting, but improves the quality of underground premises in sense of the increase of their stability, that is the reduction of need for multiple adaptation, because of deformations, by which the safety of the active mines is bigger.

In some of our coal mines (in shaft "Strmosten" in RMU "Rembas" – Resavica), the experimental use of supporting technology by rockbolting was carried out, which has given the satisfactory results.

In JPPEU mines ("Lubnica", "Rembas" – shaft "Strmosten", "Soko", "Štavalj") the opening of new excavation fields is in progress, that can use the mechanized methods of excavation, which will make the need for a faster and safer construction of shaft premises. In mentioned mines there are no limitations in matters of modernized application of mechanized construction of shaft premises and the use of a modern type of support.

CONCLUSION

On basis of all that is presented in the paper one can conclude that with the existing technology of the shafts premises construction in Serbian coal mines, the insufficient construction speed is achieved, and with certain improvements of working operations the construction speed could be increased for 30%.

Because of the opening of new excavation fields and the necessity for a quicker construction of shaft premises and the increase of the construction speed for 30%, it wouldn't be enough with the existing technology of drilling and blasting method. Because of that in JPPEU mines ("Lubnica", "Rembas" – shaft "Strmosten", "Soko", "Štavalj"), it is necessary to accept new construction technologies (mechanized construction) and supporting (rockbolts or other modern type of support). Using the mechanized construction of shafts premises the construction speed would be increased for about 2,5 times in relation to the construction speed by technology of drilling and blasting method. Taking into account the experience of the experimental use of rockbolts too in a shaft "Strmosten", it can be concluded that by introduction of this type of a support, additionally, the construction speed and the stability of shaft premises would be increased, and costs of shaft premises construction would be reduced.

No matter the used technology for shaft premises construction in the coal mines of Serbia it is necessary to pay attention to: work organization, labour training, choice of the adequate equipment, as well as the working environment in which the shaft premise is constructed etc.

REFERENCES

Đukanović D.: Model optimizacije tehno-ekonomskih pokazatelja pri izradi podzemnih prostorija u rudnicima uglja Srbije. Doktorska disertacija, RGF, Beograd 2005.

Đukanović D.: Tehnologija izrade jamskih prostorija kombinovanim mašinama (sa osvrtom na mogućnost primene u rudnicima uglja Srbije). Monografija, Savez Energetičara, Beograd 2005.

Đukanović D.: Istraživanje mogućnosti primene kombinovanih mašina za izradu jamskih prostorija u rudnicima uglja Srbije. Podzemni radovi, 14, RGF, Beograd 2005.

Čokorilo V., Milisavljević V., Milanović R., Walker G.: Operational Experiences of Rockbolting in Rembas Colliery – Serbia. Roofbolting in Mining, Achen, Germany 2004.

Đukanović D., Đukić B., Sanković Ć.: Izbor kombinovane mašine za izradu podzemnih prostorija za uslove rudnika uglja sa podzemnom eksploatacijom. Podzemni radovi, 13, RGF, Beograd 2004.

International Mining Forum 2009, Sobczyk, Kicki & Saługa (eds) © 2009 Taylor & Francis Group, London, UK, ISBN 978-0-415-80428-8

Potential of Underground Exploitation of the Existing Resources in RTB Bor

Branislav Mihajlovic, Arso Kljajevic, Negovan Aksic
RBB BOR, Bor, Republic of Serbia

ABSTRACT: In this paper there has been carried out an analysis of the possibility and perspective for underground expoloitation of the existing mining resources in RTB Bor. The analysis has shown that the ore deposits "Veliki Krivelj" in Bor and "Juzni Revir" in Majdanpek, which are at present being exploitated by open-pit mining, in fact, make it possible to use the mass underground exploitation. In Bor's underground mine there is also a perspective for the ore deposit exploitation of so-called "current operations" and especially the ore body "Borska Reka", as well as the ore body "Coka Marin", located in the area of Jasikovo.

KEYWORDS: Perspective, underground exploitation, resources, bulk mining

INTRODUCTION

Mining and copper production in Bor has had a tradition of one hundred years. During that period there has been mined and processed more than seven hundred million tons of ore out of which several million tons of copper and several hundred tons of gold and silver have been produced. The same importance has had the production of precious and rare metals as well as some other products of secondary processes. The value of that production, according to the present standards amounts to approximately twenty billion US dollars. For the towns Bor and Majdanpek, as well as the wider social community, the copper and precious metal production during all this period of time, has had a great importance and it has given an enormous contribution to their development. Considerable resources, which have been gained by the profitable production in RTB Bor, were directed to the development of Bor and its wider region.

The profitable copper production in Bor have lasted until the mid eighties of the last century. It was the time when the first symptoms started indicating that the copper production in RTB Bor started lagging. Year by year since than the profitability, economy and the prosperity of business have become real problems. By the end of twenties of the last century and in the beginning of this century, there was formed a gap between the selling copper price on the world's market and its production price in RTB Bor, which together with some other accompanying causes, brought to the production decrease from 140.000 tons per year in the beginning of the nineties to the amount of 15.000 tons in recent years. Therefore, the copper production level in RTB Bor was considerably reduced by 10% of the optimal. The copper price boom on the world's market by the end of 2004 only temporarily stopped the obvious disaster of RTB Bor, but at the same time, that chance, which lasted for four years, was not taken for its consolidation and cutting down the business costs. The sudden copper price fall in the third quarter of this year caused a great imbalance between the achieved production costs and the selling copper price and so the enterprise entered a great current illiquidity with a very uncertain outcome.

The current average cost price for a tone of cathodes produced from the local concentrate amounts from 6.000 USD/t to over 7.000 USD/t, but the minimal possible production in the existing technological conditions and under the maximal optimization, realization and reduction of all possible expenses, it cannot come under 5.000 USD/t. This type of exploitation has not got any perspective based on economic bases for a longer period of time and it should be given up as soon as possible for the sake of avoiding unnecessary loses.

The ecology problems in Bor have reached the level of a working one so that the same way of production must not be continued in the future development and, what is more, special attention must be paid to solving the ecology problems. Air pollution caused by Bor's smelter facilities represents the main source of permanent danger for all citizens of Bor as well as its near and wider region. Collectors of waste water, both civic and industrial ones are in a very bad condition, some of them are broken down and watertight and some create the permanent danger an ecology disaster. The collector of Krivelj's river, under the flotation tailing "Veliki Krivelj" is in the worse condition and it is the biggest source of danger in the whole region. The village Ostrelj, located directly under it, and the village Krivelj are the most endangerous, not only in Bor's Municipality but even wider. This is the reason why RTB Bor, in its plan for further development, must have dynamicly worked out concept for solving all these problems as well as provide proper means for their solutions.

In RTB Bor, there is a development strategy in power which has been implemented in the basis of tenders for choosing a strategic partner, which is now in effect. That concept anticipates the level of investing in RTB Bor in the amount of total 300 million USA dollars, and according to its structure, it contains the investment in the mine development, financial cover for historical ecology damages and building a new smelter.

REVIEW OF EXISTING MINING RESOURCES OF RTB BOR WHICH CAN BE EXPLOITED BY UNDERGROUND MINING

RTB Bor has got at its disposal a great deal of row material, which consists of copper ore, polymetalic ore, non-metal deposits as well as technogenic row materials. By actual reports on mining reserves, the following reserves are covered:

Table 1. Row material reserves covered by the actual reports on mining reserves which are at RTB Bor's disposal

<table>
<tr><th colspan="4">Resources</th><th colspan="2">Quantity</th><th colspan="2">Cu, t</th><th colspan="2">Au, kg</th><th colspan="2">Ag, kg</th><th colspan="2">S, t</th><th>Fe_3O_4, t</th><th>Mo, t</th><th>SiO_2, t</th><th>CaO, t</th></tr>
<tr><td rowspan="5">Underground exploitation</td><td rowspan="4">Deep pit</td><td rowspan="3">Current operations</td><td>O.B. Brezanik</td><td rowspan="3">7.262.726</td><td>1,494,900</td><td rowspan="3">70.078</td><td>19,643</td><td rowspan="3">1.727</td><td>390</td><td rowspan="3">11.771</td><td>3,064</td><td rowspan="3">820.449</td><td>259,515</td><td></td><td></td><td></td><td></td></tr>
<tr><td>O.B. Tilva Rosh</td><td>2,902,675</td><td>25,079</td><td>412</td><td>3,091</td><td>310,856</td><td></td><td></td><td></td><td></td></tr>
<tr><td>O.B. P_2A</td><td>2,865,151</td><td>25,356</td><td>925</td><td>5,616</td><td>250,128</td><td></td><td></td><td></td><td></td></tr>
<tr><td colspan="2">O.B. Borska reka</td><td colspan="2">319,969,179</td><td colspan="2">1,599,846</td><td colspan="2">65,274</td><td colspan="2">518,350</td><td colspan="2">24,957,596</td><td></td><td>11,455</td><td></td><td></td></tr>
<tr><td>Coka Marin</td><td colspan="2">O.B. Coka Marin I</td><td colspan="2">220,714</td><td colspan="2">4,767</td><td colspan="2">1,307</td><td colspan="2">8,945</td><td colspan="2">40,435</td><td></td><td></td><td></td><td></td></tr>
<tr><td rowspan="3">Open-pit</td><td colspan="3">Veliki Krivelj</td><td colspan="2">465,150,392</td><td colspan="2">1,511,726</td><td colspan="2">31,584</td><td colspan="2">180,478</td><td colspan="2">24,390,328</td><td></td><td></td><td></td><td></td></tr>
<tr><td colspan="3">Juzni revir</td><td colspan="2">246,082,778</td><td colspan="2">881,904</td><td colspan="2">46,654</td><td colspan="2">353,684</td><td colspan="2">3,102,988</td><td></td><td></td><td></td><td></td></tr>
<tr><td colspan="3">Cerovo*</td><td colspan="2">170,000,000*</td><td colspan="2">595,000*</td><td colspan="2">11,900*</td><td colspan="2">170,000*</td><td colspan="2"></td><td></td><td></td><td></td><td></td></tr>
<tr><td rowspan="2">Non-metals</td><td colspan="3">Belorecki pescar</td><td colspan="2">23,566,924</td><td colspan="2"></td><td colspan="2"></td><td colspan="2"></td><td colspan="2"></td><td></td><td></td><td>21,641,825</td><td></td></tr>
<tr><td colspan="3">Zagradje</td><td colspan="2">14,454,940</td><td colspan="2"></td><td colspan="2"></td><td colspan="2"></td><td colspan="2"></td><td></td><td></td><td></td><td>7,844,695</td></tr>
<tr><td>Tehnog. raw</td><td colspan="3">Smelting slag</td><td colspan="2">11,197,587</td><td colspan="2">80,062</td><td colspan="2">3,158</td><td colspan="2">50,389</td><td colspan="2">68,417</td><td>962,992</td><td>4,625</td><td></td><td></td></tr>
<tr><td colspan="4">TOTAL RTB BOR</td><td colspan="2">1,257,905,240</td><td colspan="2">4,743,383</td><td colspan="2">161,604</td><td colspan="2">1,243,228</td><td colspan="2">53,380,263</td><td>962,992</td><td>16,080</td><td>21,641,825</td><td>7,844,695</td></tr>
<tr><td colspan="18">* Potentiality
O.B. - Ore body</td></tr>
</table>

It must be said right now that after the mine accident, which had happened in Bor's deep pit on 19 February 2007, the level of the exploitating reserves, so-called "current operation" was reduced to about a million tons in the ore body "Brezonik". The reserves of the ore bodies "Tilva Ros" and "P_2A", after the exploitation closing project has been finished, will probably be erased in them. There is a little probability that part of these reserves, relating to the mentioned project, will stay in exploitation. It means that the so-called "current operation" has been cut led down to only a few years of exploitation of the ore body "Brezonik", and it is only after that the deep pit can, if nothing has been undertaken, in the meantime, can be left without any prepared reserves for mining, which, in a word, would mean its inevitable closing. For that reason, it is important to make a serious analysis on the possibilities of developing the new exploitating fields in the deep pit Bor.

ORE BODY "BORSKA REKA" ABOVE THE XVII HORIZON

The basic problem the deep pit faced in the period after the accident on 19 February 2007, was the impossibility to achieve the optimal production capacity, that is, the quantity of 150.000 t/month, the minimum of 100.000 t/month. Since the production was carried out only in the ore body "Brezonik", the quantity of 20.000 t/month represents the current limit of the deep pit. There should be considered the fact if it is possible to provide a short-time production increasement in the deep pit to the optimum. In that sense, there has been done the "Feasibility study for exploitation of copper ore from the "Borska Reka" deposit in the I stage accompanied by saving the terrain surface". This study has shown that in part of the ore body "Borska Reka" above the XVII horizon, which by its all characteristics technically belongs to the so-called "current operation", can very easily and quickly, with very low investments, be included in the production of the following ore quantities in the exploitating field:

Table 2. Ore reserves of the exploiting block of ore body "Borska Reka" above the XVII horizon

Ore	Cu	Cu in ore	Au	Au in ore	Ag	Ag in ore	Mo	Mo in ore
t	%	t	g/t	kg	g/t	kg	g/t	t
7.965.814	0,46	36.557,25	0,195	1.552,36	1,673	13.326,28	48	328,577

The problem lies in a very low copper ore content of 0,46% Cu, that is 0,506% Cu in the equivalent and that is why, besides all economy measures taken out of that exploiting block, it is impossible to achieve any lower production price of 5.000 USD/t copper in concentrate. Obviously, it is not attractive at the moment.

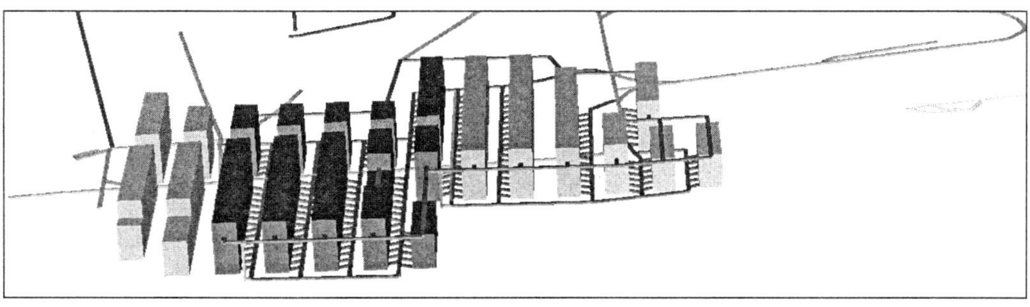

Figure 1. The exploiting block of ore body "Borska Reka" above XVII horizon

However, the study also showed that, by increasing the deep pit production to the quantity of 100.000 t/month, the expenses were distributed according to sections, that is, the greatest part of the deep pit costs has been transferred from the ore body "Brezonik" to the ore body "Borska Reka", so that it turned out that, under that conditions, the copper price in concentrate, produced from the ore belonging to the ore body "Brezonik", falls down to the level of around 3.000 USD/t.

It means that the study generally has shown that, by the exploitation of the ore body "Borska Reka", it is not possible to achieve good economy effects for that section, but on the other side, for the sake of increasing the deep pit production to its optimum, the total mining effects have been considerably improved.

ORE BODY "D" UNDER XIII HORIZON

The ore body "D" was mined up to the XIII horizon. It belongs to a group of ore bodies "Tilva Mika". The area in the mentioned location has been mined several times, but not completely. The research was carried out by drifts and prospect drillings on the IX, XI and XIII horizons. Recently, in the period from 1994. to 1995, the research was carried out from the horizon XIII by prospecting drilling of the deeper parts of the ore body. The prospecting drilling from the level XIII by prospecting drilling of the deeper parts of the ore body. By prospecting drilling from the drift 11–78, which was carried out in 1975. the data about ore body inclination under the XIII horizon were obtained. From the drift 11–78, the fan of 6 drill holes downwards were drilled. The following drill holes were drilled: B-228–233/XI. From these five drill holes, the deeper parts of the ore body were drilled obtaining the drill holes B-229 and 230. At that time, it was found out that the ore body was divided into two parts. The distance between them was 40–60 m. The research from the XIII horizon was carried out on five profiles, by drilling horizontal and slanting drill holes downwards in 1994. and 1995. By these researches, the following reserves were balanced in the area of XIII–XV level:

Table 3. Up-today explored potential ore reserves of the ore body "D" under the XIII horizon

Ore, t	Cu, %	S, %	Au, gr/t	Ag, gr/t
2.230.000	0,92	14,42	0,56	1,60

The space under the XIII horizon, in the area between the ore bodies D, D_1 and C, P_2A up to the Bor's fault, represents a very interesting and high-potential space for exploration. The grounds for working out the research program have been made and the decision on starting operations is being expected. In case the explorations have been approved, the positive results are anticipated. Expectations are to increase greatly the raw material bases of the deep pit for the quantities of over million tons of high-quality ore, which will be economically approximately on the ore level of the ore body "Brezonik".

ORE BODY "BORSKA REKA" UNDER THE XII HORIZON

By the report on ore reserves there has been balanced the ore reserves of the upper part of the ore body "Borska Reka" up to L–450 m:

Table 4. Balanced ore reserves of the ore body "Borska Reka" up to the level L–450 m

Ore	Cu	Cu in ore	Au	Au in ore	Ag	Ag in ore	Mo	Mo in ore
t	%	t	g/t	kg	g/t	kg	g/t	t
319.969.179	0,5	1.599.846	0,204	65.273	1.62	518.350	35,8	11.455

The level L–175 m marks the boundary under which there is a considerable enrichment of Cu grade as well as some accompanying metals in the ore. On the side of the hoisting shaft of the deep pit Bor, there was installed part of transporting-hoisting system across the level L–235 m. According to the concept, it was imagined that this room should be prolonged horizontally, up and down slope in the length of 800 m up to the ore body "Borska Reka", without disturbing the current production. In that way it would be possible to take a considerable part of reserves of the ore body "Borska reka" and include it into the production. Unlike the reserves above the level L–155 m, these reserves are not possible to include into production easily with law investments.

The second option of exploiting that part of the ore body "Borska Reka", has been worked out in the study, which had been done by the World Bank in 2005: "The analysis of a potential application of a block-caving method in the deposit "Borska Reka". In that analysis, the exploiting block for this method application was defined from:

Table 5. Exploiting reserves of the exploiting block of the ore body "Borska Reka" up to the level L–450 m, according to thc concept of the study of the World Bank by the method "Block caving"

Ore	Cu	Cu in ore	Au	Au in ore	Ag	Ag in ore	Mo	Mo in ore
t	%	t	g/t	kg	g/t	kg	g/t	t
137.430.000	0,56	765.438	0,23	31.746	1.68	231.020	45	6.198

The exploitation, according to this option, can be taken into consideration only after the open pit mining of the ore body "Veliki Krivelj" has been finished, since, according to the scenario, worked out in the mentioned dtudy, the ore treatment has been anticipated to be done in the Flotation "Veliki Krivelj".

Figure 2. Isometric display of the exploiting block of the ore body "Borska Reka" according to the scenario of the World Bank study by the method "Block caving"

It is also interesting to mention that the price cost of the copper concentrate, which would be produced according to the scenario, worked out in the mentioned study, is 1.600 USD/t. On one side, it is positively the cheapest copper which could be produced in RTB Bor, and on the other side, because of the precondition that the mine "Veliki Krivelj" would stop working at long date-which means it can not be short-term but long-term and at the same-time consequently and exceptionally the stable perspective for RTB Bor.

ORE BODY "COKA MARIN"

The ore body "Coka Marin" is located within the area of the village Jasikovo, some fifteen kilometers away from Majdanpek. The complete explorations have been carried out, the whole report on mining reserves has been done, as well as the mining project and at the moment it has been waiting for obtaining the permission for the operations to start on this region. This ore body can be exploited only by underground mining. For the sake of technological explorations as well as the potential exploitation, there has been worked out the mining rooms in the total length of 1.400 m with two levels (horizons) and the connection between them has also been done.

By this report, the following reserves have been covered:

Table 6. Ore reserves of the ore body "Coka Marin" covered by the report

Ore	Cu	Cu in ore	Au	Au in ore	Ag	Ag in ore	S	S in ore
t	%	t	g/t	kg	g/t	kg	%	t
220.714	2,16	4.767	5,92	1.307	18.32	8.945	18,32	40.435

The ore body can be easily included into the exploitation with a very low level of investment, providing the problems of ore processing and ecology have been solved positively.

ORE DEPOSITS CURRENTLY IN EXPLOITATION BY AN OPEN PIT MINING ("VELIKI KRIVELJ" IN BOR AND "JUZNI REVIR" IN MAJDANPEK)

In RTB Bor, at the moment, the ore bodies "Veliki Krivelj" in Bor and "Juzni Revir" in Majdanpek have been exploiting by the open pit mining, while the exploitation of the ore bodies "Cerovo" in Bor and "Severni Revir" in Majdanpek have been ceased.

Out of the four mentioned ore deposits, which have been exploited by the open-pit mining so far, the ore body "Severni Revir" has not got the geology interpretation in the software programme "Gemcom", while the remaining three ones have. That very fact excludes this ore body from the possibility to take it into account for both open-pit and underground mining. Because of the lack of this geology block-model, it is not possible to provide the accurate determination of the optimal exploiting block and it is both the basis as well as the condition for further techno-economical explorations.

The ore body "Cerovo" has got the geology block-model in the software programme "Gemcom" but for some other unfavorable characteristics, it has also been excluded as a possibility for a profitable underground mining as it is not possible in the existing conditions to define the technological and economical profitable solution for ore exploitation by underground mining from the already mentioned ore body.

Finally, the conclusion turn out to be that, under the existing conditions and on the base of the available data, the exploration on the possibility to switch to the profitable undermine mining, is only possible in ore bodies "Veliki krivelj" and "Juzni Revir". To the previous considerations we should add a very favorable fact for using the block caving method, and it is that both ore bodies, according to their formation, are almost ideally vertical, with regular vertical distribution of mases and metal content. That characteristics showed that it is possible to define the qualitative exploiting blocks in both ore bodies.

Table 7. Geology ore reserves of ore deposits which are being exploited at the moment by an open-pit mining method, where the possibilities of underground mining have been considered

Ore deposit	Wet ore t	Dry ore t	Cu dry ore %	Cu t	Ag dry ore g/t	Ag kg	Au dry ore g/t	Au kg
Juzni revir	460,286,476	446,477,882	0.316	1,410,870	1.100	491,126	0.168	75,008
Veliki krivelj	568,827,298	551,762,479	0.340	1,875,992	0.400	220,705	0.070	38,623
In total potentialy	1,029,113,774	998,240,361	0.329	3,286,862	0.713	711,831	0.114	113,631

The total quantities of geology mining reserves in ore bodies "Veliki Krivelj" and "Juzni Revir", which had been covered by exploring the possibility for economic underground mining as basis for defining the exploiting blocks, are given on Table 7.

In the text ahead, we will talk in more details about the way we define the exploiting blocks for mining these two ore bodies by the block caving method.

The existing geology block models for ore bodies "Juzni Revir" and "Veliki Krivelj", have been done by respecting already adopted parametêrs for their up to date open-pit exploitation. The dimensions of "solids" are 15×15×15 m, because of the height of both open pits – which is 15 m. Their height arrangements overlap with the levels of working floors (benches) on both open pits. The geology block-model has taken into consideration the "solids" from DGS 0,2% Cu onwards.

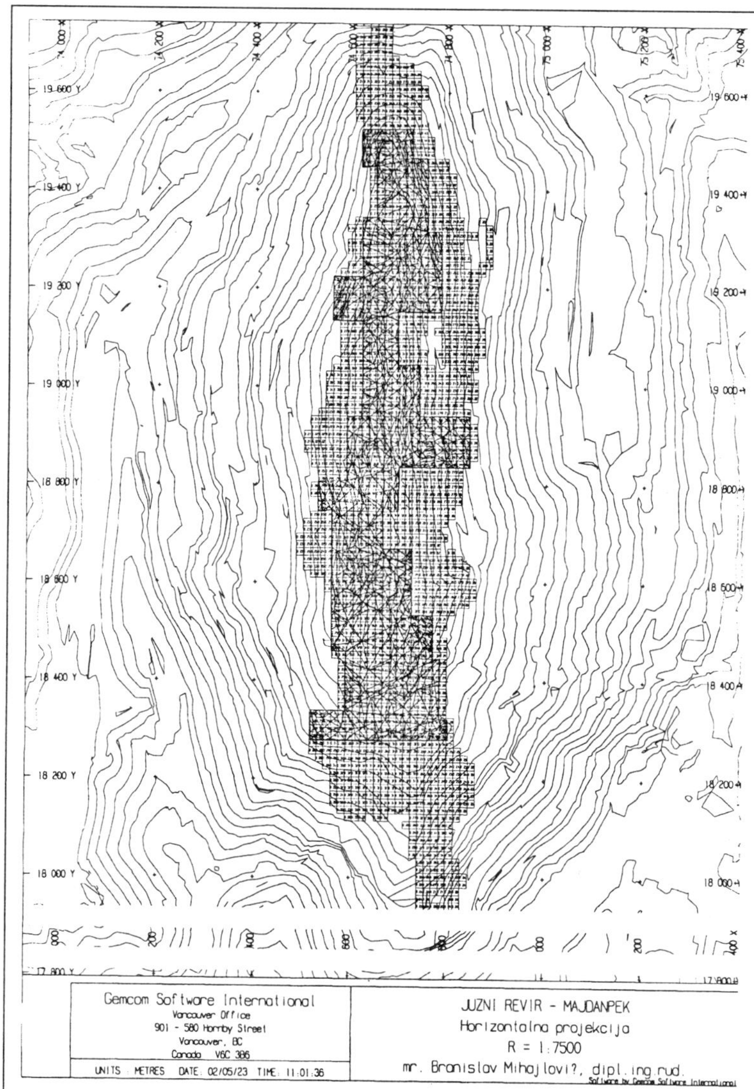

Figure 3. Defined exploiting block of the ore deposit "Juzni Revir" in comparison to the block-model and the open pit contour

While defining the dimensions of exploiting block for the bottom of both exploiting blocks in the first step, it has been taken the level L–100 m, as the lowest exploiting one and in the block-model shown level. From the top side, the limit of exploiting blocks marked the existing contours of the open pits, within the limits of which are the existing bottoms on the levels of L+125 m on the open pit "Juzni Revir" and L+85 m on the open pit "Veliki Krivelj".

Then, in all horintal intersections (on each 15 m) from the block-model are "thrown away" solids with the copper grade over 0,2% Cu and under 0,3% Cu. So worked out horisonal intersections were on all levels overlapped one over the other and that surface, was enclosed within which the solids were predominant on all levels with the copper grade of over 0,3% Cu.

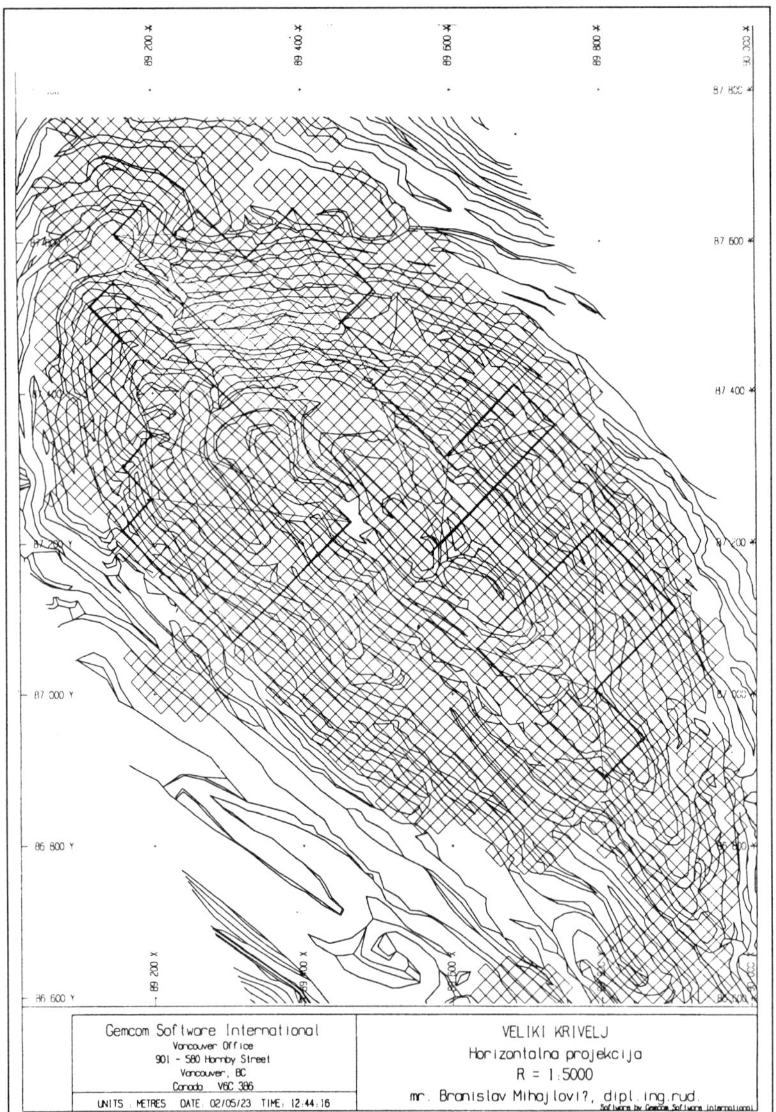

Figure 4. Defined exploiting block of the ore deposit "Veliki Krivelj" in comparison to the block-model and the open pit contour

The blocks of so defined exploiting blocks present vertical closed surfaces (areas) which connect the block bottoms with contoures of open mines. So, the exploiting blocks have been obtained for both ore bodies with the dominant presence of solids having the grades of over 0,30% Cu, whose bottom is on the level L–100 m, and with vertical blocks and the top part which represent the existing contours of open pits. In such defined contours of the exploiting block, "Gemcom" is to calculate the following parameters:

- average grade of metals: copper, gold and silver,
- average bulk density of ore mass,
- volume of the exploiting block,
- quantity of taken ore,
- quantity of taken copper,
- quantity of taken gold, and
- quantity of taken silver.

In the further procedure, firstly, the analysis of the copper distribution with increasing depth has been done. Then, it was noticed at the ore body "Juzni Revir" that the grade of copper and precious metals decreases with the depth, and at the ore body "Veliki Krivelj" it increases.

On the base of the previously noticed characteristics, we have come to the conclusion that in further work on defining the exploiting blocks, for the sake of obtaining better ore quality in the blocks, there should be carried out the "shortening" of blocks in their heights, but in such a way that the exploiting block of the ore deposit "Juzni Revir" should be shortened upwards, and the ore deposit "Veliki Krivelj" downwards. Further calculations of some basic parameters, in such defined exploiting blocks, have already proved their positive effects. At the ore body "Juzni Revir", the floor (bottom) of the exploiting block was gradually lifted by 15 m, first on level L–85 m, then on level L–70 m and so on. It also meant that the average copper grade gradually rose but the quantity of the exploiting reserves fell down. The definite determination of the bottom limit of the exploiting block was done by defining the optimal relationship of the average copper grade in it, the quantity of taken exploiting reserves and the supposed quantity of investing capital into the development and preparation for mining. The optimal relation of the mentioned parameters has been achieved when the bottom of the exploiting block was lifted to the level L+35 m.

Table 8. Exploitation reserves of ore deposits "Juzni Revir" and "Veliki Krivelj" given by exploitation blocks

Exploiting block		Volume m^3	Specific weight t/m^3	Ore quantity t	Grade Cu %	Quantity Cu t	Grade Au g/t	Quantity Au kg	Grade Ag g/t	Quantity Ag t
Veliki Krivelj	Block 1	15,540,000	2.5	38,850,000	0.4185	162,578	0.07	2,720	0.39	15,152
	Block 2	17,600,000	2.499	43,982,400	0.4419	194,358	0.07	3,079	0.39	17,153
	Block 3	8,858,000	2.5	21,462,500	0.4407	94,585	0.07	1,502	0.39	8,370
	In Total	41,725,000	2.499	104,270,775	0.4329	451,388	0.07	7,299	0.39	40,666
Juzni revir		28,055,000	2.653	73,924,925	0.4282	316,457	0.253	18,733	1.508	111,508
Total		69,780,000	2.554	178,195,700	0.431	767,935	0.146	26,032	0.854	152,174

As for ore deposit "Veliki Krivelj", the previously mentioned procedure of iteration could not after satisfactory results, that is, it was not possible to achieve a suitable average copper grade in the exploiting block (it has been achieved the maximum of slightly over 0,38% Cu), while the quantity of exploitation reserves was regularly greater than required for a fifteen-year operation, and that was defined as an optimal lifetime of exploitation. For that reason it was decided that at the ore deposit "Veliki Krivelj", one more increase of the lower limit grade to be done-from 0,30% Cu to 0,35% Cu. On the horizontal profiles, all solids were thrown out, having the copper grade over 0,30% Cu as well as under 0,35% Cu. So a new exploiting block was determined consisting of three blocks separated by ore of poorer quality from 0,35% Cu. The bottom of the exploiting block remained on the level L–100 m, while the upper surface (area) was limited by the existing contour of the open pit, but only to the bench on the level L+110 m, that is, the level of the terrain L+125 m. The ore above the level L+125 m was thrown out from the exploiting block for the sake of a poor Cu grade. Interpretations of, in that way defined exploiting block, are given on Figure 3 (page 9) and Figure 4 (page10).

The calculation of the covered reserves in the exploiting blocks of ore deposits of "Juzni Revir" and "Veliki Krivelj", is given in the Table 8.

ANALYSIS OF POTENTIAL UNDERGROUND MINING OF EXISTING MINING RESOURCES IN RTB BOR

The previous overview of resources, which can be exploited by underground mining in RTB Bor, points to the fact that there are different levels of potentials of their exploitation. That is why they can be classified into several groups of their potentials:

1. The resources of low potentials. These are the potentials which can not be exploited if the price cost is lower than 4.000 to 5.000 USD/t of copper in concentrate. These include the resources, so-called of the "current operations" of Bor's deep pit, that is the ore body "Brezonik", the ore body "Borska Reka" above the XVII horizon and the ore body "D" under the XIII horizon. These also include the ore body "Juzni Revir" in Majdanpek, as it has not got enough quantity of ore nor the proper height of the exploiting block for a successful method of "Block caving".
2. The resources of medium potentials. The only one is the ore body "Coka Marin". Although the ore of this resource is rich, the doubts about its optimal way of processing as well as it efficient ecology protection have not been resolved yet, both in its exploitation and processing. That is the reason why the potentials of this resource has been put into the function of the mentioned problems.
3. The resources of high potentials. These cover the ore body "Borska Reka" under the XVII horizon of the deep pit of Bor and the ore body "Veliki Krivelj". The ore body "Borska Reka" represents the resource of RTB Bor of the highest level of potentials. It can be exploited in the period of at least thirty years, in the quantity of 8 to 9 million tons of ore, that is 25.000 to 30.000 t of copper per year by the block caving method and with the price cost of 1.600 USD/t of concentrate. The ore body "Veliki Krivelj" can also be exploited, even more successfully than the ore body "Borska Reka", by the block caving method, but the parameters of such a method are economically worse than the parameters which can be obtained by exploiting the ore body "Borska Reka". Unfortunately, both resources can not be of a short-term but of a long-term potentials of RTB Bor.

SUMMARY

The underground mining of the ore resources of RTB Bor, located in the Bor's deep pit ("Jama"), can not in a short-term period provide the copper production according to the lowest price of 4.000 to 6.000 USD/t of concentrate, and such production can exist in the forthcoming period from five to ten years, depending on the fact if there would be any investments into it. These investments would refer to the introduction of exploitation of the ore body "Borska Reka" above the XVII horizon as well as the ore body "D" under XIII horizon. The introduction into exploitation of the ore body "Coka Marin" would, if the problems of ore processing and ecology protection have been solved successfully, considerably improve the total economical effects of the underground mining in RTB Bor. If investments have not been provided, than in the next three-four years only the ore body "Brezonik" will be exploiting, and the price cost will be from 8.000 to 10.000 USD/t of concentrate, and consequently, the deep pit ("Jama") will have to be closed.

Contrary to the previous facts, a long-term perspective of underground mining in RTB Bor is exceptional. All that only under the condition of providing considerable investments. The investment into the most potential resource of RTB Bor, the ore body "Borska Reka", according to the scenario of the "World Bank", and its rough calculations, would amount to around 170 million USD. From the moment of making a decision on investment up to the beginning of the exploitation, it will take at least five, and in practice it usually takes seven years, then the period of increasing the production from the initial level up to the full capacity, will take three to five years, and in total, it will take some ten years. Since the ore body "Borska Reka" is located into the Bor's deep pit ("Jama"), this deep pit should not be closed because its vital infrastructure would be used for the needs of its future exploitation. It means that the Bor's deep pit should be overviewed with more complexity, taking into account its continuity. If a decision has been made to exploit the ore body "Borska Reka", and it is the only proper decision, than there must be provided the continuity of exploiting the so-called "current operations".

Finally, it is necessary to mention that mining is a human activity which has always been carried out under a hard pressure. The skillfulness and successfulness of people, who are dealing with mining, is being measured by their ability, diligence and knowledge to solve these problems. In solving the problems relating to a successful development of mining in Bor in the forthcoming period, there must be introduced essential changes, all in accordance with the advanced trends. Which have already been achieved in the world's mining science of RTB Bor, and which could be carried out only by teams of the most professional and able people.

REFERENCES

The World Bank. Assessment of RTB Bor and Analysis of Reconstruction Options/Privatization. January 2005.

The World Bank. Analysis of Potentiality for Applying a Block Caving Method in the Ore Deposit "Borska Reka". May 2005.

RGF Belgrade. Feasibility Study of a Short-Term Investment (Period from 2006 to 2011) in Copper Concentration Production in RTB Bor. September, Belgrade 2005.

Mihajlovic B.: Exploration of the Possibility for Economic Underground Mining of the Existing Copper Ore Reserves. PhD Dissertation, Bor 2003.

Mihajlovic B.: Potentials of Copper Production by Underground Mining from Bor's Deposit. Round Table, Brestovacka spa near Bor, July 2006.

International Mining Forum 2009, Sobczyk, Kicki & Saługa (eds) © 2009 Taylor & Francis Group, London, UK, ISBN 978-0-415-80428-8

Research Results on Possibilities for Development and Mining of "Borska Reka" Ore Body in Bor Ore Deposit

Zivorad Milicevic, Vitomir Milic, Igor Svrkota
University of Belgrade, Technical Faculty in Bor, Republic of Serbia

ABSTRACT: Massive copper ore deposits occur in the Eastern Serbia. Bor ore deposit, with many ore bodies, has been exploited for 105 years. Underground mining has been present all the time, while surface mining lasted for about 80 years. Extraction of active ore bodies is coming to an end. In the mean time, ore body "Borska Reka" has been explored. This ore body is situated beneath lowest levels in Jama Bor underground mine and it is so massive that it could carry out the future of copper ore mining in Bor ore deposit. Some considerations on extraction of ore from this ore body are given in this paper.

1. INTRODUCTION

Some 100 million years ago, in upper chalk geological period, at the area of Eastern Serbia, there were two volcanic phases that brought out Timok Igneous Complex. Hornblend – biotitic andesine, along with pyroclastic rock was dominant in this complex [1]. During this period, copper mineralization was created, and most significant were following deposits: Bor, Veliki Krivelj, Cerovo and Majdanpek (Fig. 1). Bor ore deposit, as the most important one, was discovered at the beginning of 20th century. Year of 1903 is considered as the beginning of copper ore mining in Bor area. Many years later, in 1961, Majdanpek ore deposit was activated, while Veliki Krivelj open pit started in 1983. However, most of the deposits today are considered as almost exhausted, unless price of copper at the World market increases rapidly and stays at high level.

Majdanpek, Cerovo and Veliki Krivelj copper ore deposits were extracted by surface mining and these open pits almost reached designed final depth. Their further extraction requires extension of open pits, which means serious investments in removing of burden. In a period of huge increase of copper price, it seemed possible. However, price went down again, thus making it very questionable.

In Bor ore deposit, surface mining was definitely ended in 1993, while underground mining is still present. After the end of surface mining, sublevel caving methods were introduced in underground mining. Their application was possible only when open pit became inactive. Explorations of deeper parts of this area lead to discovery of massive ore body, "Borska Reka", named by river in southwestern part of the deposit, bellow which the ore body was found (Fig. 2). This ore body occurs slightly bellow the sea level, and also bellow the Horizon XV, the lowest level of Jama Bor underground mine, at K.–75 m.

Top part of Bor ore deposit reaches 360–500 m above sea level, while service shaft nearby stands at K.+440 m.

Final depth of "Borska reka" ore body has not been determined yet, since the deepest exploration drill holes reached 1,200 m depth (Fig. 3), which means that ore body is at least 800 m deep. Ore body thickness varies. At K.–400 m level, thickness reaches 500 m, and average thickness is about 300 m. These dimensions are referred to contour of ore body with minimum 0.3% Cu, with some 600 Mt of ore, with average content reaching 0.6% Cu.

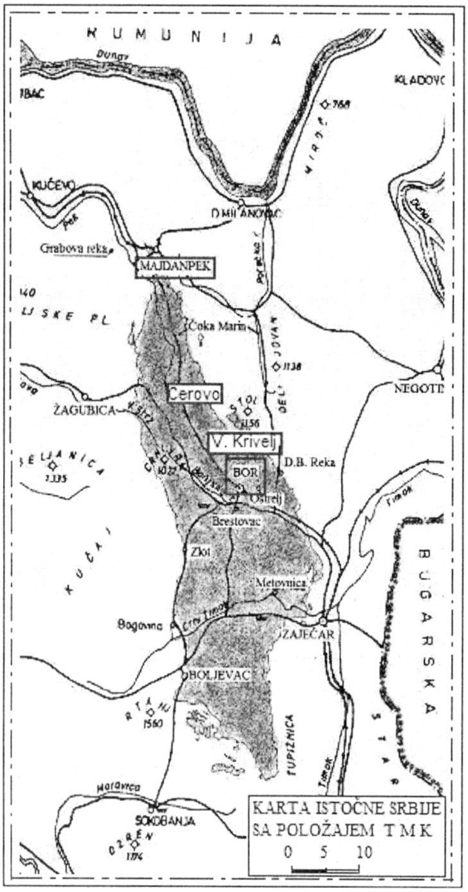

Figure 1. Timok Igneous Rock Complex in Eastern Serbia

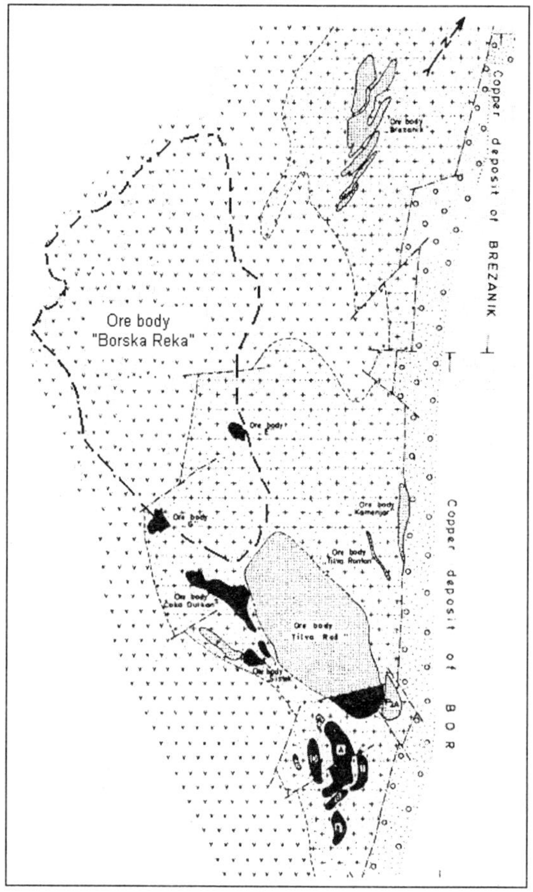

Figure 2. Layout of "Borska Reka" ore body

Beyond any doubt, this is the largest ore body in this hydro-thermally altered area. However, some of the following facts have to be considered:

- ore body depth requires high costs of development and extraction;
- average content of copper in the ore is considered as very low for underground mining;
- there are many important structural and construction objects at the ground surface above the ore body (Fig. 4), which exclude possibility of application of caving methods, unless the objects were displaced before mining.

In such situation, numerous researches were focused on optimal way of extraction of this ore body. However, after 30 years, we still don't have final decision about its extraction.

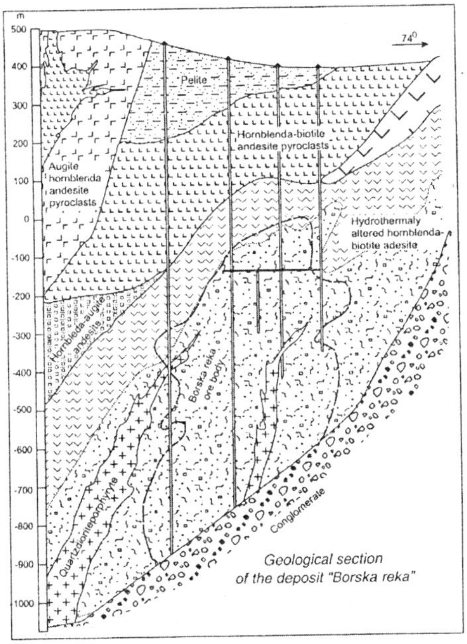

Figure 3. Geological cross-section of ore body "Borska Reka" with exploration drill holes

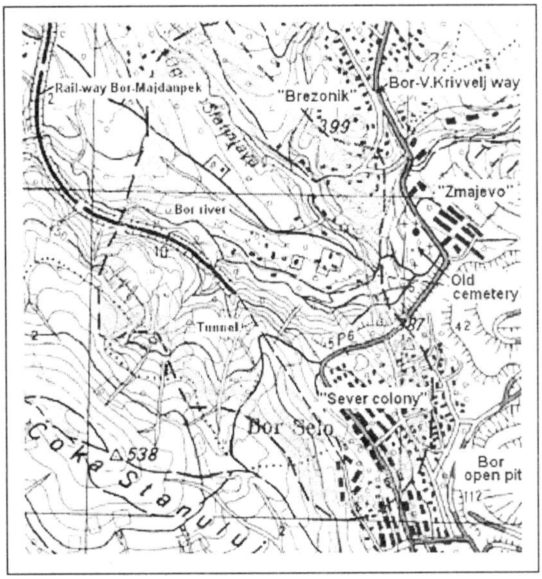

Figure 4. Ground surface above ore body

2. CONSIDERATIONS AND DECISIONS IN ORE BODY DEVELOPMENT

Theoretical researches of possibilities for economically justified extraction of this ore body started in early eighties of the last century. Many studies and mine designs were presented, but none of them was acceptable. Overall conditions were always unfavorable. Low metal content and low price of copper disabled any attempt to begin the extraction process. Rapid increase of metal prices enabled serious plans far beginning of extraction. However, even in such favorable period, that goal was difficult to reach due to residual difficulties in former years. At the same time, the process of restructuring of RTB Bor Company was in progress. Unfortunately, several attempts of Company privatization failed, thus making the overall situation even more difficult. Besides, serious decrease of metal prices added more difficulties. Any way, any price of copper bellow 3,000 USD/t limit, disables the extraction.

There are two main problems that have to be solved in case of activation of "Borska Reka":

- mine output is limited by capacity of mineral processing plant in Bor;
- mining method has to be designed in order to match all the requirements, and some of the most important constraints are ore body depth and objects at the surface.

Ore body development should be done in two phases. Phase one includes mining with annual outputs up to 4 Mt, and ore processing in Bor processing plant. Phase two, with outputs over 4 Mt and up to 10 Mt of ore, includes Veliki Krivelj processing plant, which has more capacity.

Transition to lower parts of the deposit in Jama Bor underground mine goes gradually. First concentration level was horizon XIII at K.–16 m. This was main transport level until excavation in Tilva Ros ore body reached horizon XIII. As the excavation went deeper, so did the development. A slope drift from horizon XV, at level K.–75 m, was driven down to horizon XVII, at level K.–155 m (Fig. 5). Exploration drifts at horizon XVII have proved the existence of new ore body, "Borska Reka," at that level.

Based on theoretical researches, further development was designed as it is shown bellow (Fig. 5):
- deeper parts of ore deposit are developed by slopes;
- main hoisting slope in the southern part of the deposit, GIN-21/-235, with a belt conveyer, was driven to enable development of horizon XIX, at level K.–235 m, which is currently the lowest level of the mine;
- main transport drift, GTH-235, with a belt conveyer, connects GIN and central ore pass (CRO) – this way the ore is hauled from CRO to hoisting shaft bunker, at horizon XIII;
- horizon XVII, at level K.–75 m and horizon XIX, at K.–235 m, are connected by central ore pass (CRO) and service-ventilation shaft (PVO);
- primary crusher is installed in the central ore pass and it is used to comminute the ore coming from upper parts of the deposit.

This way new concentration level, horizon XIX at level K.–235 m, was developed and previous concentration level became inactive. Scheme of development is shown in Figure 5.

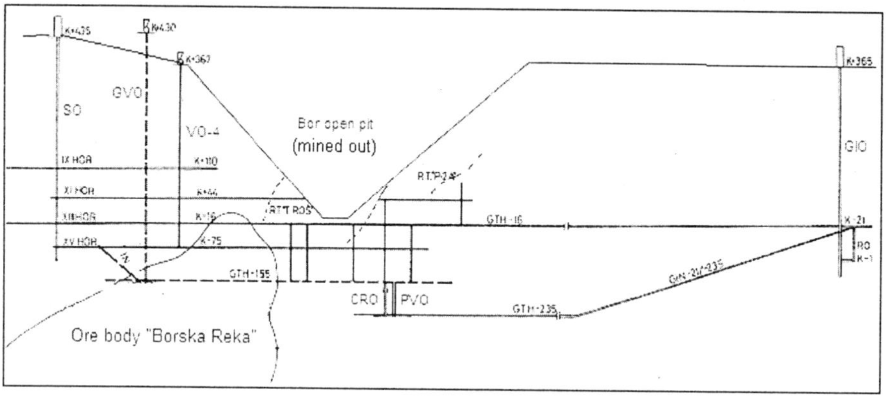

Figure 5. Scheme of Jama Bor underground mine: GIO – hoisting shaft, SO – service shaft, VO-4 – ventilation shaft, GVO – new ventilation shaft, not driven yet

Concept of mining in "Borska Reka" ore body predicts gradual increase of output, with development in phases. Development should be carried out through transport and service slopes. Drifting of slopes should also advance in succession with ore body excavation. One of possible designs, made by Institute of Mining and Metallurgy, is shown in Figure 6. This variant of ore body development includes new hoisting slope towards Veliki Krivelj mineral processing plant, since this plant has more capacity than the old one in Bor. Expected decrease of production in Veliki Krivelj open pit in the future, would make the plant available for processing of ore from underground mine.

3. RESEARCHES OF MINING METHODS

Considerations on selection of mining method were guided by both properties of rock in the ore body and situation at ground surface. Since "Borska Reka" is a massive ore body, first considerations were focused on caving mining methods, especially block methods (Fig. 7). These methods provide high outputs and low costs, which is crucial for mining of low-graded ore. At least five or six variants of induced block caving were considered, including:
- induced block caving with at least one sublevel drift inside 80 m level;
- induced block caving with long drillholes, drilled from a single level;
- semi-level induced block caving, with mining in two independent levels with spacing matching half of level height.

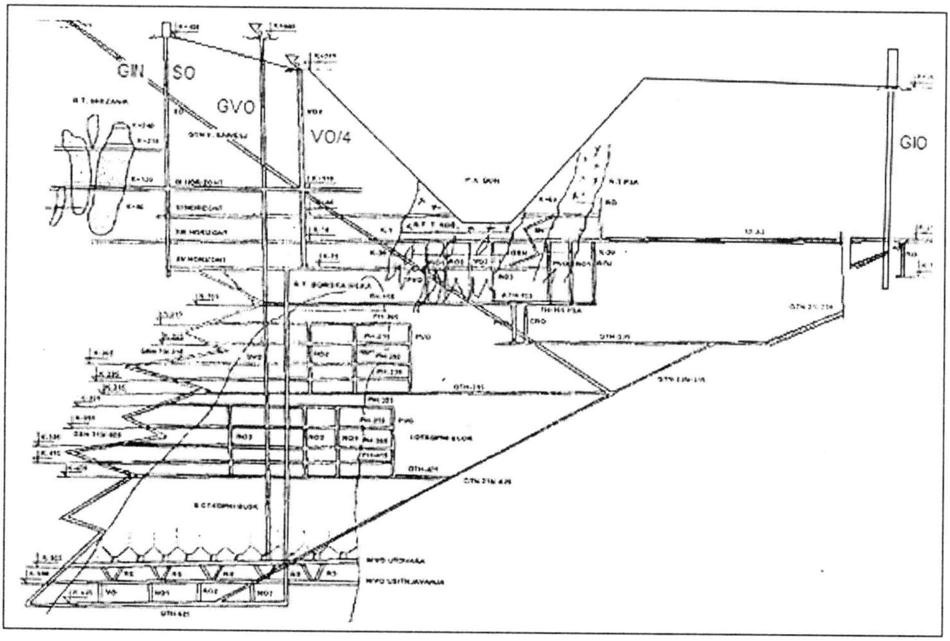

Figure 6. Variant of development of "Borska Reka" ore body with new hoisting slope (GIN) towards Veliki Krivelj mineral processing plant

Beside these variants, methods with mining in large blocks were considered, too. Large blocks are both very high and wide, and the ore is gained in large zones. This way bulk of ore is gained at once, thus providing decrease of mining costs.

This variant is specific due to increased spacing between block bottom and transport level, thus providing more volume of ore in the ore passes. Transport drift of one level will be used as sub-level drilling drift in the lower level.

This way drilling drifts will be driven from transport drift, and there will be 2 or 3 of them in ore body due to its thickness.

Each level (transport level, loading level and drilling level) has independent ventilation system, thus avoiding more ventilation shafts in the ore body. In order to decrease the depth of drillholes, undercut level is lifted up to 30 m. Blasting is performed in large zones (15–20 m).

Block caving methods were excluded from considerations at that time, regardless on a fact that they have the lowest costs. The reasons for their exclusion were ore body depth and rock properties.

However, considerations of block caving and possibilities for its application are still present, especially because they were seriously researched in eventual transition of currently active open pits to underground mining.

Finally, considering the fact that there are numerous civil engineering objects at the surface above "Borska Reka" ore body, cut and fill mining methods also became interesting. Most important objects at the surface are following:

- railroad tunnel,
- regional road,
- Borska Reka river,
- old cemetery,
- parts of urban areas.

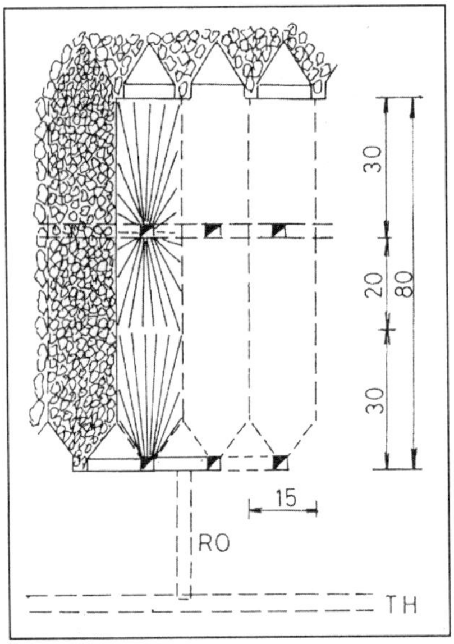

Figure 7. One of the variants of induced block caving

Estimated costs of displacing those objects reach 30 million US dollars, and such a huge investment would have to be realized before the beginning of mining.

So, the researches went further. Block methods with open stopes would preserve the surface, while later open stopes could be filled with floatation waste. Disadvantage of this group of methods is poor ore recovery, way under 50%. On the other hand, for such a huge ore body, with enormous ore reserves, costs of objects displacement reach only 0.06 US$ per ton of ore, thus making caving methods more interesting.

New idea came along in last couple of years. In order to avoid huge investments at the beginning of mining, and save time needed for displacement of objects at the surface, ore body summit could be mined in large rooms, while caving of pillars would be induced after several years, when objects are already moved. After that, mining would continue with caving methods. This idea has also been considered only theoretically, as a study.

More details on some variants of mining methods will be given in next paper.

CONCLUSION

"Borska Reka" ore body, with its enormous ore reserves, is the only alternative for long-term mining in this area. All of the remaining ore bodies were already mined out, or reached the final phases of extraction. The ore is low-graded, so it requires increase of copper price to provide economically justified production.

Many comprehensive researches were performed during the years and many designs of ore body development and mining were offered. Some solutions for development are already applied in Jama Bor mine.

Difficult economic situation of the mine, caused by many political and economic turbulences in last two decades, disabled faster development of the mine since many of suggested solutions wait for more convenient economic situation. Hopefully, the improvement should be provided through the process of privatization and establishing of strategic partnership with some respectable mining company. If copper price remains relatively stable, we expect intensification of design and preparation works and final reviving of underground mining in Bor ore deposit.

REFERENCES

Drovenik M.: Origin of Bor and Other Copper Deposits in Its Surroundings (East Serbia). SAZU, Ljubljana 2005.
Janković S.: Ore Deposits of Serbia. (In Serbian). Republican Fund for Geological Researches and RGF, Belgrade 1990.
Milićević Ž.: Sublevel Caving and Induced Block Caving. (In Serbian). TF Bor 2008.
Technical Documentation of Institute of Mining and Metallurgy and Technical Faculty in Bor.

International Mining Forum 2009, Sobczyk, Kicki & Saługa (eds) © 2009 Taylor & Francis Group, London, UK, ISBN 978-0-415-80428-8

Mining Productive Systems Organization for Optimal Ventilation and Air Supply

Miodrag Miljković
RBB BOR, Bor, Republic of Serbia
e-mail: mailto:val.miljkovic@ptt.rs

Predrag Golubović
RBB BOR, Bor, Republic of Serbia
e-mail: mailto:predraggolubovic@yahoo.com

ABSTRACT: Costs for mine ventilation and air supply depend of natural conditions of layer's deposits, selected technology and techniques for layer access and mine production system organization (construction). In this article is discussed organization of production system for optimal air supply and mine ventilation.

Optimal mine supply with air and mine ventilation is accomplished based on optimization of equivalent hole, which is function of all influencing factors. Based on that, optimal cross sectional size of mine's rooms (average values) is determined, which has to be accomplished before selection of mining equipment for exploitation to fit-in their dimensions in to the dimensions of rooms' cross sections.

KEYWORDS: Air quantity, equivalent hole, ventilation costs

INTRODUCTION

Ventilation in mine provides necessary air quantity for work of people and machines, comfortable climate parameters and aeration, de-concentration under MAQ and removal of harmful materials from the working environment atmosphere in the form of gases, vapours, dust and smoke. Ventilation costs are calculated in relation of the tons of produced ore and they are obtained from derivate and simplified relations for average conditions:

$$C_v = \left(N_v T C_e\right) / Q_{god}, \quad C_v = 22{,}2 Q^3 C_e / A^2 Q_{god}, \quad \mathrm{din/t} \tag{1}$$

where: $N_V = 1{,}25RQ^3/(1000\eta)$ [kW] – installed power of main fan motor; $R = \alpha UL/S^3$ – aerodynamic resistance of mine rooms [Ns^2/m^6]; η – ventilation plant efficiency ($\eta = 0{,}7–0{,}8$); Q – necessary air quantity for optimal production system ventilation [m^3/s]; $T = 365 \cdot 24$ – continual work of fan per year [h]; Ce – costs for electrical energy [din/kWh]; Q_{god} – annual ores production in mine [t]; $A = 1{,}19\sqrt{R}$ – equivalent mine opening ($R = 1{,}42/A^2$) [m^2]; $\alpha = \lambda\rho/8$ – coefficient of room's aerodynamic resistance; U – rooms circumference expressed by cross sectional area of the room S [m^2], $U = o\sqrt{S}, o \approx 4$ [m]; o – coefficient for determination of room's circumference in function of S – – room's cross-sectional area (o = 4, for round cross-section; o = 3,54 for trapezoidal cross-section and 0 = 3,9 for arched cross-section).

Costs for mine ventilation are proportional to the third power of the air quantity Q, used for mine ventilation, and in inverse proportion to the square of equivalent mine opening, at constant price for electrical energy and constant annual production of the mine. Value of equivalent mine opening, (A), includes systematic data of mine design (cross-section of mine rooms, coarseness, branches, etc.). Electrical energy costs and annual mine production are values which are not dependable of technical solution for ventilation of mine and they are constant for any way of underground production system arrangement.

By Regulations of technical norms for underground exploitation of coal, metal and non-metal ores it is defined that air quantity for mine ventilation should be determined in accordance with all influencing aspects on working conditions and selected the most suitable for all prescribed conditions. From aspect of ventilation costs, selected air quantity should not be optimal, but it is necessary to provide safety work.

Suitable working conditions in production system can be achieved by use of other protection solutions, making that with technical organization of production system an optimization of mine supply with air and ventilation with their optimal (minimal) costs could be accomplished.

SELECTION OF OPTIMAL AIR QUANTITY FOR VENTILATION OF PRODUCTION SYSTEMS IN SERBIAN MINES

Necessary air quantity for mine ventilation is determined based on the following parameters:

- Gas-carrying ability of coal layer and surrounding rocks, including uneven occurrence of the dangerous gas;
- Gases at blasting;
- Gases from inner combustion motors, where minimal air quantity must not be under 4 (m^3/min/ /kW);
- Minimal air streaming velocities of 0,25 [m/s];
- Climate working conditions;
- Number of mounted fans for separate ventilation;
- Dusting of air;
- Number of workers at the face of work and in;
- Lowest atmospheric pressure.

Necessary air quantity for mine ventilation is determined to provide safety work, and for these purposes it could be the highest quantity obtained from counted parameters. From the aspects of the safety work and ventilation costs, selected air quantity could be optimal if safety work conditions are provided by application of additional technical measures. Usually the biggest air quantity is necessary to provide suitable working climate in the production system or for de-concentration and removal of chemical harmful material from working environment. If for comfortable working conditions, the additional technical measures are applied (air cooling and/or control of emissions of harmful materials), than it is possible to decrease necessary air quantities for mine ventilation related with this demands.

DETERMINATION OF OPTIMAL AIR QUANTITY TO PROVIDE COMFORTABLE CLIMATE IN PRODUCTION SYSTEM

To obtain optimal climate for work it is necessary to provide sufficient fresh air quantity from the surface. Necessary air quantity for ventilation of mining production systems depends of branching (size) of production system, used working equipment and geothermal changes in mineral ore's layer. Based on demonstrated natural factors and applied technical solutions on production system, working conditions could be forecasted in relation of climate parameters and necessary quantity of fresh air from the surface could be determined to provide suitable climate parameters on the working faces. If determined air quantity could be provided in the mine rooms (with air stream velocities less than 8 m/s), than such solution could be accepted as rational. In relation with this solution arise a question of determination of production system depth with ventilation with air form the surface without its cooling. Descending into lower depths at some production systems suitable working climate should be provided by cooling systems application.

In mining rooms air gets wormer due to auto compression and air circulation, rocks geothermal heat, machine work, people work, vaporization of hot waters, expansion of comprised air, etc. Because of that for each mining room, separate ventilated ore body or ventilation branch in mine with compound ventilation schemes, heat balance should be made based on sum of heat from all sources of heat. Heat balance is described in details in further text.

Based on heat balance for the room, temperature increment in room and necessary quantity of fresh air to provide suitable climate working conditions is determined.

Temperature increment of air in room or pit, under condition that relative humidity of air is the same, could be calculated for each heat source separate according to the relation:

$$\Delta t = \frac{Q_i}{c_p \cdot Q_v \cdot \rho_s}, \ ^\circ\text{C}$$

where: Q_i – quantity of heat developed in room or pit; c_p – specific air heat at constant pressure (1,006 kJ/kg K); Q_V – air flow; ρ_S – Average air density, calculated as arithmetic mean of inlet air and air on the working site or according to the following relation:

$$\rho_s = \frac{b - \varphi \cdot E}{R_{vaz} \cdot T} + \frac{\varphi \cdot E}{R_{vp} \cdot T}$$

where: b – barometer pressure on the working site [mbar]; φ – relative air humidity at working site [%] (in coal pit it is around 90%); E – partial pressure of saturated water vapour at temperature t_S [mbar]; R_{vaz} – gaseous air constant (287,1 J/kg K); R_{vp} – gaseous constant of the water vapour (461,5 J/kg K); T – temperature of air [K].

Air quantity to provide suitable working conditions, regulated by technical norms for underground exploitation of metallic and non-metallic mineral ores or regulations for underground exploitation of coal could be determined according to the formulae:

$$Q_v = \frac{\sum Q_j}{\rho_s \cdot (I_{konf} - I_{ulaz})}, \ \text{m}^3/\text{s}$$

where: ΣQ_i – heat balance for rooms for preparation and layer stoping works; I_{konf} – air enthalpy for scaled climate parameters and atmospheric pressure at working sites, calculated according following formulae:

$$I_{konf} = c_p \cdot t_{S2} + c_{vp} \cdot x_2 \cdot t_{S2} + c \cdot x_2$$

t_{S2} – air temperature on dry thermometer according to the regulations; c_{vp} – specific heat of the water vapour (1,86 kJ/kg k); x_2 – humidity content in the air at working site, calculated according known formulae:

$$x_2 = 0{,}622 \cdot \frac{\varphi \cdot E}{b - \varphi \cdot E}, \ \text{kg/kg}$$

c – evaporating heat at 0°C (2500 kJ/kg °C); I_{ulaz} – enthalpy of the inlet air, calculated according to the formulae:

$$I_{ulaz} = c_p \cdot t_1 + c_{vp} \cdot x_1 \cdot t_1 + c \cdot x_1$$

t_1 – temperature of the inlet air (variable during the day, month or year) calculated according to the formulae:

$$t_1 = t_0 + r \cdot H$$

t_0 – average annual temperature in the mine area; r – air temperature increment due to adiabatic air compression in summer and winter period (0,004–0,001); H – eye depth, x_1 – water vapour content in the inlet air for the conditions t_1 and air humidity $\varphi = 90\%$.

Norms for climate parameters are given in the Table 1.

Table 1. Scale of norms for climate parameters in mine

Minimal air velocity, m/s	Air temperature t_S in °C for relative air humidity φ, %		
	60–75	75–90	more than 90
0,25	24	23	22
0,50	25	24	23
1,00	26	24	24
2,00	26	26	25

If air velocity in rooms of inlet and outlet air stream is bigger than 8 m/s, at in-taking of such calculated air quantity, than on such way scaled climate parameters could not be provided, but cooling devices for air cooling must be mounted on the working sites. By introducing the cooling devices it would not be necessary to intake previously calculated air quantity, but another, considerable smaller.

NECESSARY AIR QUANTITY FOR VENTILATION OF DEEP MINES

At calculations of necessary air quantity for ventilation of deep mines, starting point is that mine air is heated due to cooling of the rock massifs, under condition to eliminate other heat sources in horizontal mine rooms. Heat quantity taken by air is:

$$Q_H = K_\tau \cdot U \cdot L \cdot (t_{SS} - t_{VS}), \text{ kJ/h}$$

And air temperature at entrance of the mining rooms is:

$$t_2 = t_S - (t_S - t_1) \cdot e^{-\frac{K_\tau \cdot U \cdot L}{Q_V \cdot \rho_S \cdot c_p}}$$

where: K_τ – coefficient of unstationary heat change between rocks and air; U – volume of the mine room; L – length of the mine room; t_{SS} – average rock's temperature in the massif, calculated according to:

$$t_{SS} = \frac{t_S + t_{VS}}{2}$$

t_S – rock's temperature at layer's depth H, calculated according to:

$$t_S = t_0 + \frac{H - h_0}{G}$$

t_0 – temperature of neutral layer; H – layer's depth; h_0 – depth of neutral layer (≈ 20 m); G – geothermal degree; t_{VS} – average temperature of air in the room, calculated according to the formulae:

$$t_{VS} = \frac{t_1 + t_k}{2}$$

t_l – temperature of the inlet air at pit eye ($t_l \approx t_0$); t_k – comfortable temperature, equal to the temperature of the room's outlet air t_2.

By values substitution for $t_2 = t_k$, formulae becomes:

$$\frac{t_S - t_k}{t_S - t_1} = e^{-\frac{K_\tau \cdot U \cdot L}{Q_v \cdot \rho_S \cdot c_p}}$$

Taking that:

$$-\frac{K_\tau \cdot U \cdot L}{Q_v \cdot \rho_S \cdot c_p} = -X$$

than:

$$e^{-X} = \frac{t_S - t_k}{t_S - t_0}$$

By determination of e^{-X} from the expression $(t_s - t_k)/(t_s - t_0)$, value of the parameter X can be calculated.

Finding the solution for Q_v from previous formulae, expression for determination of necessary air quantity to provide comfortable working climate in mining rooms becomes:

$$Q_v = \frac{K_\tau \cdot U \cdot L}{3600 \cdot \rho_S \cdot c_p \cdot X}, \quad m^3/s$$

To use this formula it is necessary to determine the coefficient of unstationary heat exchange between rocks and air. Determination of this coefficient is very complicated. One of the formulas for its determination, for conditions in wet rooms and water evaporating from the room walls, has the following expression:

$$K_\tau = \frac{0{,}8 \cdot \lambda^{0{,}8} \cdot \left(c_S \cdot \gamma^{0{,}2}\right)}{d_{ekv}^{\ 0{,}6} \cdot \tau^{0{,}2}}, \quad kJ/hm^2\ °C$$

where: λ – heat conductivity [kJ/hm °C]; c_S – rocks' specific heat of rocks [kJ/kg °C]; γ – rocks density [kg/m^3]; d_{ekv} – equivalent room diameter [m]; calculated according to:

$$d_{ekv} = 0{,}96 \cdot \sqrt{S} = \frac{4 \cdot S}{U}$$

S – room's cross-section [m^2]; τ – working hours from opening to the room's use [h].

In Table 2 are shown thermal characteristics of some rocks in copper ore's mines.

Table 2. Thermal characteristics of some rocks

Rock	Density, kg/m^3	Temperature conductivity, α 10–4, m^2/h	Specific heat, c_S, kJ/kg °C	Heat conductivity, λ, kJ/hm °C
Quartz, Albeit, Porfiryte	2851	39,60	0,78	8,78
Chalcopyrite	4716	37,26	0,86	15,04
Pyrite	4620	35,78	0,90	23,00
Sand	2440	43,70	0,85	9,20

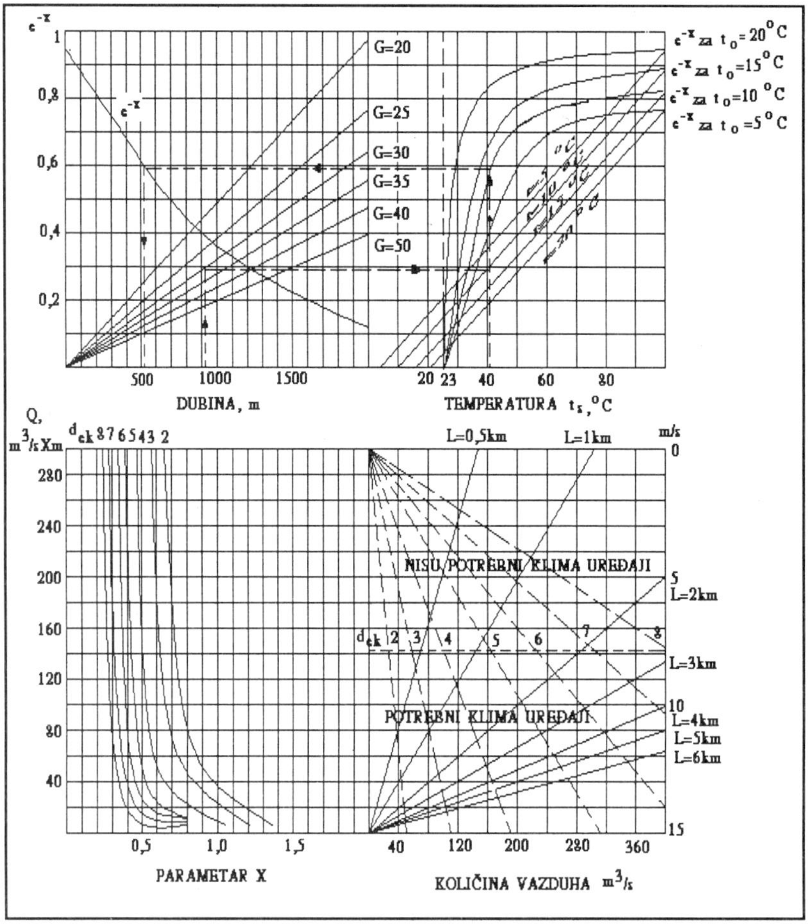

Figure 1. Nomogram for determination of necessary air quantity for deep pits ventilation

Necessary air quantity for ventilation of deep pits could be calculated using nomograms shown on Figure 1. As it is obvious from the monogram, the most important parameters from formulae for determination of necessary air quantity are taken into consideration (parameter X, mine depth, geothermal degree, rock temperature, equivalent room's diameter, room's length, average annual temperature of the inlet air). By following the arrows of the algorithm for determination of necessary air quantity for ventilation of deep production systems, the fourth field of the monogram is reached where according to the rooms' length and their cross-section en evaluation of the necessary air quantity for ventilation. Taking into consideration of inlet rooms cross-section and allowed air flow velocity of 8 m/s in them, it is evaluated that quantity of air should be decreased. This quantity could be decreased by inlet air cooling. Based on given monogram it is possible to evaluate is artificial cooling of inlet air into production system necessary or not. When is, for providing of comfortable climate conditions on the working sites, necessary to intake great amount of air, and due to room's cross-section realized air flow velocity is bigger than 8 m/s, cooling of the inlet air is needed in the aim to decrease its quantity necessary to realize flow velocities less than 8 m/s.

Other demands for air quantities for production system's ventilation could be discussed in the scope of optimal design of production system and selection of optimal technology, technical and organization of work:

- Optimal solution for providing of suitable climate in production system is found by comparison of costs when it is accomplished by taking in of great air quantity into production system or when devices for acclimatization and aeration are built in production system, accompanied with providing of optimal quantity of air for ventilation.
- Demand for diluting of gases below MAQ and their removal from working site, could be achieved by taking in the necessary air quantity, obtained by comparison of ventilation costs, increased air quality and costs for previous degassing of the layer through systems for degassing.

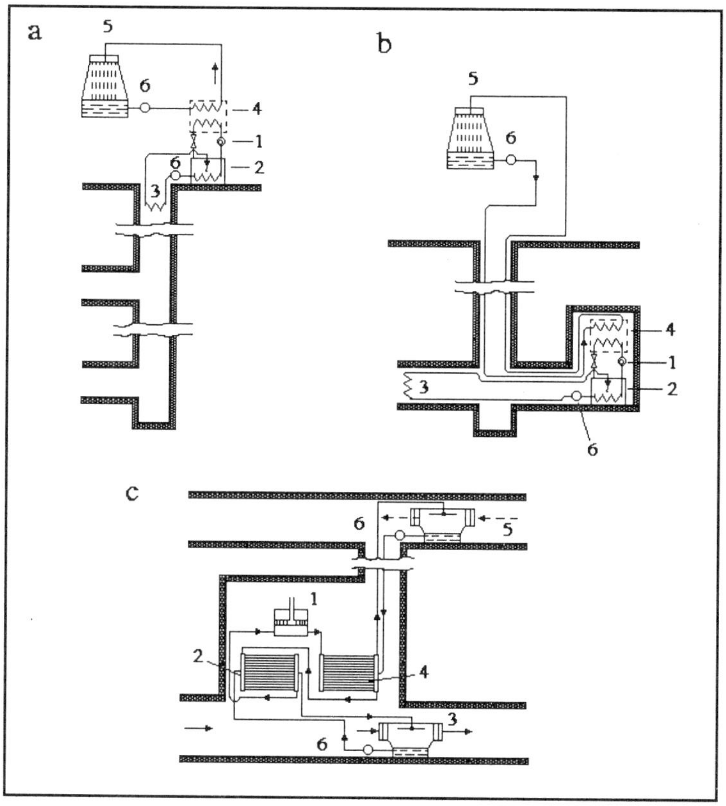

Figure 2. Schemes for pit air cooling: a – cooling of the total air stream; b – air cooling at deep horizon with condenser at the surface; c – cooling of the inlet air at the working site with taking the heat into the outlet air stream of the horizon. Marks: 1 – compressor, 2 – evaporiser (cooling of the coolant medium), 3 – air cooling device, 4 – condenser, 5 – cooling water device mounted on the surface or in the pit, 6 – circulating pumps for coolant circulation

Optimal air quantity for dust diluting is needed at application of technical measures for control of production and removal of dust into the working sites of production systems using wet and dry procedures and by comparison of their costs and costs for ventilation of mines without special protection application.

Demand for great amount of air for gases and dust diluting after massive blasting at working sites could be decreased by better organization of blasting in the periods when there are no workers in mine, by extension of ventilation time and by direction of greater amount of air into the working site where blasting has been performed, using flow regulator.

Optimization costs for exploitation of ores using equipment driven by different energy sources is accomplished by variant observation of total exploitation costs, costs for development of necessary mining rooms profiles, together with costs for mine ventilation and selection of optimal variant for stoping method. Together with this an optimal ventilation air quantity is determined.

SELECTION OF OPTIMAL CROSS-SECTION OF OPENING ROOMS

Reduction of total costs for ventilation in mines could be accomplished by reduction of aerodynemic resistance of the opening rooms (R). It is proportional to the coefficient of aerodynamic resistance α (it depends of room walls coarseness, volume (U) and rooms length (L)), and inverse proportional to the third power of the room's cross-section, ($R = \alpha UL/S^3$).

Optimization of cross-section of the opening rooms from the aspect of ventilation is accomplished based on minimal ventilation costs, i.e. air flow through them, $C_V = f(S)$ and building costs $C_{IZ} = f(S)$.

$$F = C_v + C_{iz}, \text{ din/t} \tag{3}$$

$$C_v = 1{,}25\alpha o L Q^3 \cdot 24 \cdot 365/(1000\eta S^{2,5} Q_{god})$$

or

$$C_v = B_1/S^{2,5} \tag{4}$$

$$C_{iz} = kLS/Q_r, \text{ din/t; or } C_{iz} = B_2 S \tag{5}$$

where are, beside known designations: k – costs for rooms opening and insurance [din/m^3]; Q_{god} – – annual production of [t]; Q_r – total amount of ore opened by the room [t]; L – rooms length [m].

$$B_1 = 1{,}25\alpha o 24 \cdot 365/1000\eta Q_{god};\ B_2 = kL/Q_r$$

Minimum of function $F = (B_1/S^{2,5}) + B_2S$ is on the point where its first derivation is equal zero and second derivation bigger than zero. Extreme values of this function are obtained by solution of the equation:

$$\frac{\partial F}{\partial S} = \frac{-2{,}5B_1}{S^{3,5}} + B_2 = 0, \text{ i.e. } F' B_2 S^{3,5} - 2{,}5B_1 = 0$$

$$S = \sqrt[3,5]{2{,}5B_1/B_2} \tag{6}$$

Second derivation of the function is less than zero for all values $S > 0, F'' = \partial F/\partial S$.

$F'' = -3{,}5B_2S^{2,5}$ – which means that it is minimum of the function F.

Graphical determination of optimal cross-section of the opening rooms could be accomplished by construction of functions' graphs: $C_V = B_1/S^{2,5}$, $C_{IZ} = B_2S$ and $F = B_1/S^{2,5} + B_2S$.

F_{min} is the point on the curve F where it has minimum value, i.e. where it is touched by tangent parallel to the $\tilde{S}$ – axe, as shown on Figure 3.

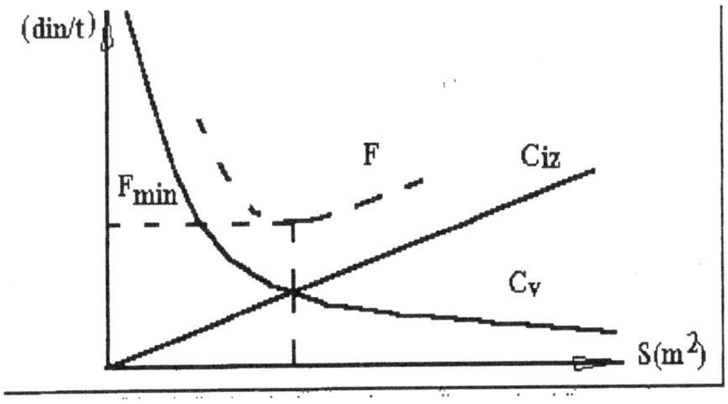

Figure 3. Graphical determination of optimal cross-section of the opening rooms

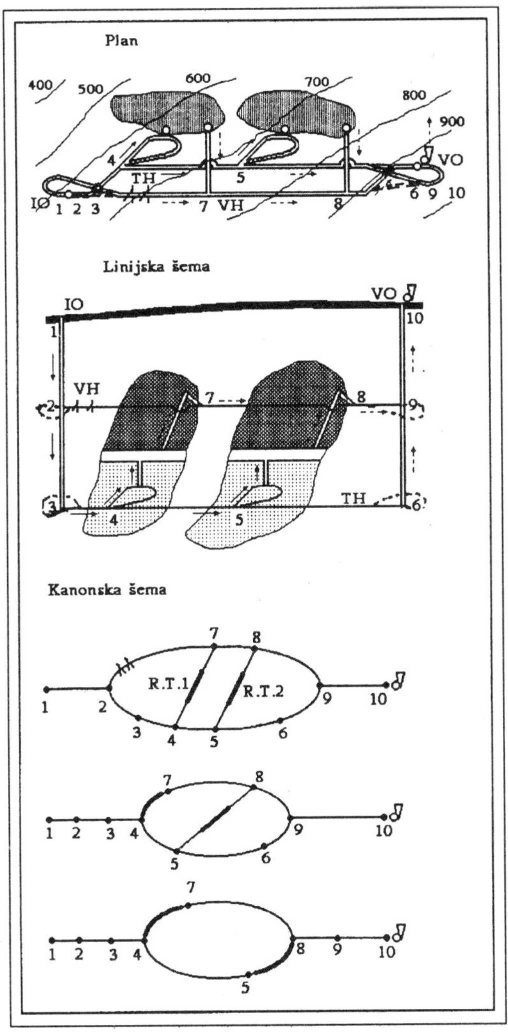

Figure 4. Graphical documentation for determination of total resistances and equivalent pit aperture

Such determined cross-sectional area of the opening rooms should satisfy the condition that air flow velocity in them is in accordance with regulations for allowed speeds and not bigger than 8 m/s, more exactly 10m/s if rooms are not intended for people transition.

Depending of selected stoping method, layer's working out and arranging of the production system is accomplished, together with air distribution. Cross-section areas of the opening rooms and preparation of layer are determined according to the dimensions of working equipment and JUS standards, and total resistance (R) of the ventilation system in mine depends of rooms cross-section area, supports in them and air distribution schemes (number of branching).Total resistance of the air distribution scheme (R), could be substituted by equivalent aperture $A = 1{,}19\sqrt{R}$. Costs for mine ventilation at constant air quantity inlet Q, is inverse proportional to the square of the equivalent aperture of the ventilation scheme, A^2 as in formulae (10), (11), (1). With bigger number of mine rooms branching, to taking in fresh air to the each working site, equivalent aperture of the production system increases and total aerodynamic resistance (R), is reduced, together with ventilation costs. Based on optimal arrangement of the production system, optimal equivalent aperture of the mine and rooms cross-section area could be obtained.

OPTIMAL EQUIVALENT APERTURE OF THE MINE AND ROOMS DIAMETER

Optimal equivalent aperture of the mine ventilation scheme and minimal opening rooms' cross-section could be determined by finding the minimum of the function of ventilation costs for the worked-out part of the layer and costs for the opening and preparation of layer for stoping, with providing of flow able stope ventilation. For this task starting point is the condition:

$$F_{min} = C_v + C_{iz}, \quad C_v = f(A) \text{ and } C_{iz} = f(A)$$

$$C_v = 1{,}25 \cdot 1{,}42 \cdot Q^3 \cdot 365 \cdot 24 \cdot C_e / (1000 \eta Q_{god} A^2)$$

or

$$C_v = B_1 / A^2, \text{din/t} \tag{7}$$

$$C_{iz} - ILS/Q_r; \; S = f(A); \; C_{iz} = IaL^{0,4} K_p A^{0,8}$$

or

$$C_{iz} = B_2 A^{0,8}, \text{din/t} \tag{8}$$

where:C_{iz} – Total costs for opening and preparation of the most distant part of the layer reduced per tone of ore which will be obtained from the preparated part, expressed in function of equivalent aperture of the ventilation scheme [din/t]; Q_r – quantity of ore which will be obtained from the prepared part of the layer [t]; A – Equivalent aperture of the ventilation scheme (at rooms connected in series, without branching, it could be determined according to the formulae: $A_{ek} = \Sigma 1/A$ and at parallel branching of the rooms it is $A_{ek} = \Sigma A_i$, where are: A_i – equivalent apertures of each room in branch and at rooms with the same A_i, $A_{ek} = nA_i$ and n – number of branches (3) [m^2]; $K_p = L/Q_R$ – – preparation norm [m'/t].

As:

$$S = f(A), \; A^2 = 1{,}42/R, \; R = \alpha UL/S^3, \; R = \alpha oL/S^{2,5}$$

follows:

$$S^{2,5} = \alpha oLA^2 / 1{,}42$$

for:

$$\alpha = 0{,}010,\ o = 4\ ,\ a = \sqrt[2,5]{\frac{\alpha o}{1{,}42}} = 0{,}24$$

and:

$$S = aL^{0,4}A^{0,8}$$

and finally:

$$C_{iz} = \frac{0{,}24IL^{1,4}}{Q_r}A^{0,8},\ C_{iz} = B_2A^{0,8},\ B_2 = 0{,}24IL^{0,4}K_p$$

I – costs for mining and insurance of 1 m of mining rooms with cross-section S [din/m^3], could be obtained from the layer's opening design, according to the valid prices for mining.

Minimum of function $F = (B_1 / A^2) + B_2A^{0,8}$ is in the point where first derivation of function is equal to zero $\partial F / \partial A = \left[(-2B_1 / A^3) + 0{,}8B_2 / A^{-0,2}\right] = 0$, followed by the equation:

$$-2B_1 + 0{,}8B_2A^{2,8} = 0$$

And solution for A is optimal:

$$A = \sqrt[2,8]{2B_1 / 0{,}8B_2}\ ,\ m^2 \tag{9}$$

For known size of optimal equivalent aperture, equivalent aerodynamic resistance $R = 1{,}42 / A^2$ could be determined. For round shapes of room's cross-sections (R) it is determined according to the formulae: $R = 64\alpha L / d^5\pi^2$ and optimal equivalent room's diameter will be, for previously adopted approximate values:

$$d = \sqrt[5]{\frac{64\alpha LA^2}{1{,}42\pi^2}} \quad \text{or} \quad d = \sqrt[5]{0{,}045LA^2},\ m \tag{10}$$

Approximate cross-sectional area for opening rooms and layer's preparation for stoping with known size of equivalent aperture, A, it would be:

$$S = d^2\pi / 4 \quad \text{or} \quad S = \frac{\pi}{4}\sqrt[2,5]{0{,}045LA^2},\ m^2 \tag{11}$$

This formula could be applied also for determination of unknown parameters in the presence of known figuration in formulae as well at active mines. It functionally connects values: L – length of the longest ventilation branch, which could be easily taken off from the ventilation plans; average room's cross-sectional area in the longest ventilation line, S, and equivalent mine aperture A, which value could be easily obtained from the fan's working regime, more exactly, Qv [m^3/s], air quantity for pit ventilation and depression Hv [Pa] of the ventilator fan sufficient to overcome the friction resistances in the mining rooms at taking in of this air quantity.

Evaluation of applied cross-section of the opening rooms and mining works could be accomplished if necessary or defined equivalent aperture of production system is introduced into the previous formulae (for coal mines with underground exploitation Ap = 1), and also the length of the longest ventilation line taken from the ventilation plan. By calculations Sp [m^2] is obtained. If pit rooms cross-section is known, S, their relation could be obtained and necessary percentage of cross-section increases.

$$100\ Sp/S = \%\ \text{increasing of existing S to the optimal or necessary Sp} \quad (12)$$

At known equivalent aperture of the production system, A, which could be obtained based on determined fan's working regime and known cross-sectional area for the opening rooms and works S, it could be checked and determine the equivalent length of the longest ventilation branch (line), Le. It could be determined from the previous formula (11). If L is explicit and known common numbers substitute with their values, it is obtained:

$$Le = S^{2,5}/0{,}0245\ A^2,\ \text{m} \quad (13)$$

Calculated unknown equivalent length Le of the longest line (hypothetical) will not match the real line length taken from the ventilation plan. Distinction $\Delta Le = Le - L$, represents increasing of line length caused by local resistances in rooms due to obstructions in rooms (frontal resistances), sudden expansions or narrow parts in the rooms, curves, etc. Big difference shows that local resistance influence is great and that the maintenance of the mine rooms is not good. This influence could be expressed in percentages according to the formulae:

$$100\ Le/L = \%\ \text{of the rooms length increasing, due to the resistance increasing} \quad (14)$$

Local resistances in rooms are reduced by branching of the air stream, regular maintenance of the room's cross-sections by removing of the materials and equipment which can block the rooms, etc. Because of that this data points to the organization of the ventilation of production system. If $Le \approx L$, than evaluation is excellent for the ventilation organization. Each increasing of the Le compared with L for 10% decreases evaluation value of maintenance and ventilation system organization for one point. That means, for the increasing of Le for 40%, the evaluation value for ventilation system organization is bad [1].

Evaluation of technical organization of ventilation system of pits could be easily accomplished by comparison of regulated value for equivalent aperture Ap for the type of production system, and real condition obtained from the fan's working regime, according to formulae:

$$A = 1{,}19\ Q_v/H^{1/2},\ \text{m}^2 \quad (15)$$

At mine pit with equivalent aperture smaller than necessary, it is needed to take measures to increase equivalent pit aperture and to get close to the optimal ventilation conditions. Relation:

$$100 Ap/A = \%\ \text{of necessary increasing of equivalent aperture}$$

Increasing of equivalent pit aperture could be accomplished by branching of the air stream i pit, reducing of the local resistances in rooms, reducing of total resistances in rooms by increasing of their cross-sections or by decreasing of aerodynamic resistance α. By shortening of the ventilation lines' length. Evaluation for pit's ventilation system organization has to be accomplished based on equivalent aperture and needed percentage of its increasing to the optimal size. If $Ap \approx A$ ventilation system state is excellent. If increasing of the aperture is necessary to obtain optimal size than evaluation of such equivalent aperture will be reduced and for each next 10% of the aperture increasing, evaluation will be decreased for one point. If needed increasing of the equivalent aperture A is 40% than ventilation organization in production system is poor [1].

CONCLUSION

Selection of the production system organization for optimal air supply and ventilation could be accomplished based on optimization of equivalent mine aperture, which is function of all influencing factors on mine ventilation. Based on optimal equivalent aperture an average optimal cross-section of opening and preparation rooms of the layer could be determined for ventilation by flowing air stream. Determination of optimal room's cross-section from the ventilation aspect should be accomplished before selection of the mining equipment, to fit the equipment dimensions into to the dimensions of the room's cross-sections. Obtained functional connections between average room's cross-sectional area, length of the ventilation's line branches and equivalent aperture make possible to analyze the state of the production system ventilation organization.

REFERENCES

[1] Miljković M.: Providing of the Necessary Oxygen Quantity for People and Machines Work in Mines. Mining and Metallurgy News, Vol. 2, Bor 1992, p. 185.

[2] Stojković Z., Miljković M., Bogdanović D.: Cross-Section Area Optimization for Ventilation Net Rooms. RGF, Proceedings, Scientific Conference: "Underground Exploitation and Aspects of Improvement and Rationalization of Technologic Processes", Belgrade 1993, p. 426.

[3] Abramov F.A.: Aero and Gases Dynamic in Mines. Nedra, Moskva.

International Mining Forum 2009, Sobczyk, Kicki & Saługa (eds) © 2009 Taylor & Francis Group, London, UK, ISBN 978-0-415-80428-8

Determination of Depth of "Borska Reka" Copper Ore Deposit at Which Air Conditioning of Working Environment Must Be Performed

Miodrag Miljković, Predrag Golubović, Dimča Jenić
RBB BOR, Bor, Republic of Serbia

ABSTRACT: The intensive technological development requires larger and larger amounts of energetic and nonenergetic mineral resources as its initial basis. Thus the need for exploiting deeper and deeper deposits arises, regarding that the shallower ones of higher profitability rate are mostly exhausted. The deep deposits exploitation generates numerous problems, or should we say challenges which must be adequately and promptly faced.

One of these problems, regarding that the rock temperature, as well as the air temperature, rises along with the exploitation depth, represents determining of depth at which air conditioning must be performed.

In this paper this problem will be analyzed on the example of "Borska Reka" copper ore deposit in RBB-Bor, Serbia.

KEYWORDS: Underground mining, deep ore deposits, ventilation, air conditioning

1. INTRODUCTION

21st century brings various challenges in mining which should be quickly and adequately answered. On account of that shallower and more profitable deposits have already been exploited, the need for exploiting deeper deposits with lower profitability rate arises. This means aggravation of working conditions in sense of longer transporting distances, higher underground pressures, longer ventilation paths and aggravated underground climate conditions. In this paper this problem will be distinguished on example of copper ore deposit "Borska Reka" in RBB-Bor, Serbia.

2. UNDERGROUND MINE CLIMATE INFLUENCING FACTORS

Underground mine climate conditions represent very complex conception, so acquaintance of all relevant factors is needed for studying them. The most relevant factors are:

1. Air temperature which is influenced by:
   - outer air temperature,
   - air warming because of air compression,
   - geothermal degree,
   - chemical processes,
   - air humidity,
   - air to wall rubbing,

   - pressured air,
   - operation of various machines and mechanisms,
   - electric devices operation,
   - excavated ore and rock cooling,
   - blasting,
   - workers influence.
2. Air humidity which is influenced by:
   - outer air temperature and humidity,
   - barometric pressure and underground mine depth,
   - surrounding rocks humidity and presence of underground water,
   - air flow quantity and speed,
   - ventilation paths length,
   - underground air temperature.
3. Air flow speed.
4. Air pressure.
5. Radiation.

3. UNDERGROUND MINE CLIMATE CONDITIONS ESTIMATION METHODS AND LEGISLATIVE

There are two methods for climate conditions estimation:

1. Separate measurement of climatic factors (temperature, humidity, air flow speed, air pressure and radiation).
2. Determination of one climatic value which represents the influence of all climatic factors (or some of climatic factors).

The first method is very complicated and unreliable because it demands a large number of instruments and very complex analysis.

Thus the method of representing of some or all relevant climatic factors by one climatic value is more accepted. This climatic value can be established by usage of special instruments or usage of special graphs and nomograms.

The most common used values are american effective temperature, belgian effective temperature and russian effective and equivalent-effective temperature.

Climatic conditions in serbian underground mining are regulated by Technical measures and underground mining working safety regulations.

The essence of these regulations is the following:

- dry thermometer temperature should not exceed 28°C and wet thermometer temperature should not exceed 23°C;
- shorten working time (six hours) for places where dry thermometer temperature is in range between 28 and 31°C, and a wet thermometer temperature is in range between 23 and 26°C. Also minimal air flow speed of 0,5 m/s is required.

4. "BORSKA REKA" ORE BODY THERMAL CHARACTERISTICS

In period between 1981 and 1985, in "Borska Reka" ore body region, thermal examination is conducted along with research drilling. The aim of this examination was to establish the intensity and disposition of thermal fields along research drilling profiles, regarding the maximum temperature values that could be expected with entering of drill-hole in an ore body. The results are presented in the table below.

Table 1. Drill-holes temperatures depending of the drill-holes depths

B-19 (1981)		B-28 (1982)		B-33 (1984)		B-39 (1984)		B-45 (1985)	
dubina, m	temp., °C	dubina, m	temp., °C	dubina, m	temp., °C	dubina m	temp., °C	dubina, m	temp., °C
1,00	14,50	8,20	8,70	19,00	14,10	100,00	15,42	69,00	13,00
7,00	14,50	13,00	9,70	36,00	14,50	150,00	15,63	117,00	14,00
9,20	13,95	19,00	10,70	90,00	15,35	200,00	16,10	148,00	15,00
23,00	13,51	23,00	11,15	114,70	16,00	250,00	17,40	189,00	16,00
82,00	15,38	31,00	11,80	129,40	16,36	300,00	18,92	226,00	17,00
93,40	15,38	53,80	12,40	146,60	16,75	350,00	20,63	250,00	18,00
112,60	16,25	73,00	11,44	173,80	17,46	400,00	21,84	287,00	19,00
203,00	18,95	165,00	13,07	194,80	19,08	450,00	23,50	325,00	20,00
228,00	19,98	179,00	13,89	231,00	19,37	500,00	24,38	360,00	21,00
352,40	23,33	209,30	15,50	252,00	20,00	550,00	25,61	406,00	22,00
371,00	23,78	209,30	15,50	322,30	21,90	600,00	26,50	449,00	23,00
406,00	24,74	219,00	16,81	360,00	22,96	647,00	27,20	485,00	24,00
429,60	25,20	223,40	17,17	368,70	23,04	703,00	28,88	539,00	25,00
432,60	25,52	229,50	17,80	425,00	24,63	764,00	29,30	585,00	26,00
472,40	26,45	251,00	18,83	444,00	25,13	801,00	30,00	628,00	27,00
493,60	26,84	283,20	20,00	504,60	26,69	850,80	30,90	682,00	28,00
501,60	27,00	283,20	20,00	540,00	27,45	896,80	31,65	736,00	29,00
513,20	27,35	317,00	21,00	592,00	28,75	953,20	32,65	781,00	30,00
536,20	27,36			649,00	29,82	993,60	33,45	833,00	31,00
546,20	28,14			691,00	30,55	1.051,00	34,59	899,00	32,00
552,00	28,15			713,00	31,00	1.141,00	35,85	968,00	33,00

Not enough measurements have been performed to statistically obtain reliable data. The measurements have been done almost immediately after drilling the holes to avoid falling down of material. In spite of that, the falling down of material occurred, making measurements at the bottoms of the drill-holes impossible. Finally, the measurements should have been repeated after a while to obtain reliable data, which was impossible.

Basing on data from the Table 1 and linear and nonlinear equations, the following values have been established:

- geothermal degree $G_s = 32{,}2°C/m$ (geothermal gradient $\sigma = 0{,}031°C/m$);
- neutral layer depth $h_{ns} = 26$ m;
- neutral layer rock temperature $t_{ns} = 11{,}5°C$ (based on Bor area average air temperature $t_{sr} = 10{,}0°C$ with temperature depending of geographic latitude $t = 1{,}5°C$ added).

5. MINE DEPTH BY WHICH AIR CONDITIONING DOESN'T NEED TO BE PERFORMED

In Bor, Serbia, Ph.D. Tomo Vukobratović B. Mining Eng. was occupied by problems considering mine climatization. He established mine depth by which air conditioning doesn't need to be performed in his doctorate [3].

The following values have been adopted for "Borska Reka" ore body:

- thermal conduction λ [W/(mK)]:
  - 0,25 for pits,
  - 2,5 for main corridors and stopes;
- with ore specific volumetric mass $\rho = 2.780$ [kg/m^3] and specific thermal capacity $c = 858$ [J/ /(kgK)], the temperature conduction $a = \lambda/c\,\rho$ [m^2/s]:
  - $0{,}104\times10^{-6}$ for pits,
  - $0{,}104\times10^{-5}$ for main corridors and stopes;
- the roughness coefficient ε:
  - 3,0 for pits and stopes,
  - 1,5 for other rooms and corridors.

The block caving method has been adopted as the excavation method that had been already successfully applied in Bor underground mine. The excavation method schematic and the linear and canonic ventilation schemes are presented on Pictures 1–3.

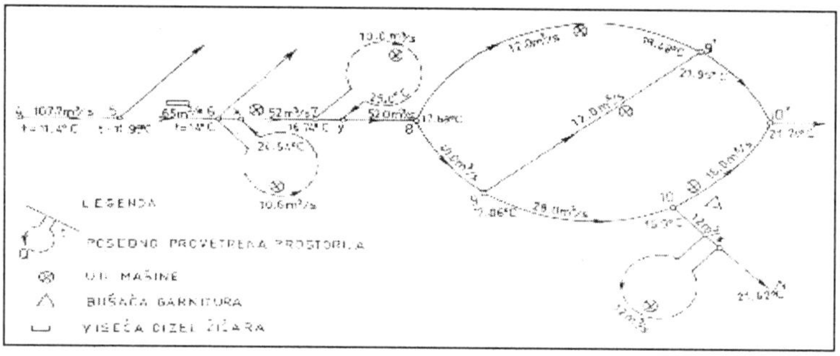

Picture 1. Canonic ventilation scheme

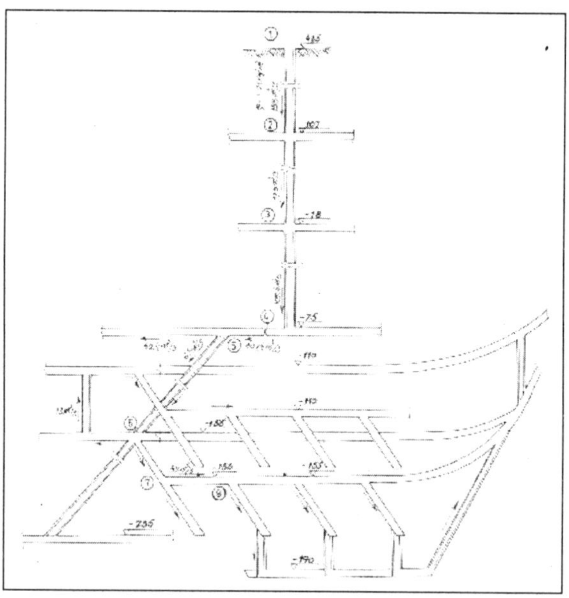

Picture 2. Linear ventilation scheme

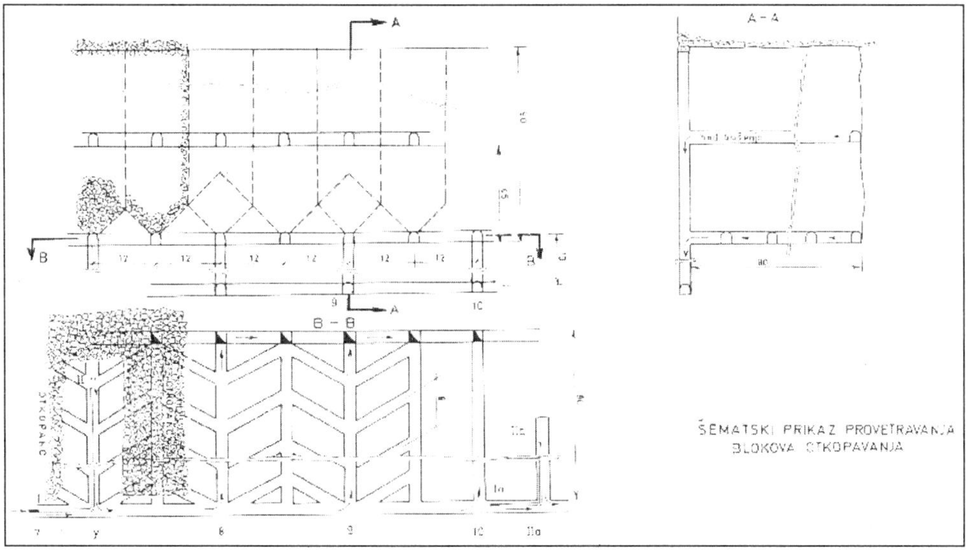

Picture 3. Excavation method schematic

On basis of upper spoken, the following formula for determination of mine depth by which air conditioning is not necessary has been constituted:

$$H_{max} = \frac{27400 - (t_{ns} - \sigma h_{ns} - 22{,}2)l_{uk} - \sigma \Sigma \Delta h_{sr.i} l_{mi} - \Sigma q}{(\sigma - 0{,}002)l_{uk} + 9{,}81} + L\sin\psi \ [m].$$

The value of depth determined this way was H_{max} = 776,0 [m] (K = 340 [m]).

6. POSSIBILITY OF THE BETTERMENT OF MINE CLIMATE CONDITIONS

The most efficient and the easiest method for betterment of the mine climate conditions is the lowering of an air temperature.

The lowering of an air temperature without use of air conditioning devices can be achieved by:

- stopping oxidation processes,
- ventilation paths insulation,
- splitting the air flow,
- increasing the amount of incoming air.

The possibilities for this betterment are not unlimited. Sometimes the air conditioning must be performed. The most common applied methods for air conditioning are:

- creating coldness on surface and air cooling on surface,
- creating coldness on surface and underground air cooling,
- underground coldness creating and underground air cooling.

The most common applied machines for air conditioning are those for underground coldness creating and underground air cooling. The schematic of such device is presented in Picture 4.

Lately applied machines are those with small dimensions and great cooling possibilities. One of those machines, for instance, is the german machine EW 1/80 which has a very small freon generating set VK 80. In Picture 5 the use of such machine is shown.

The similar machine is also used in "Pniowek" underground mine in Poland.

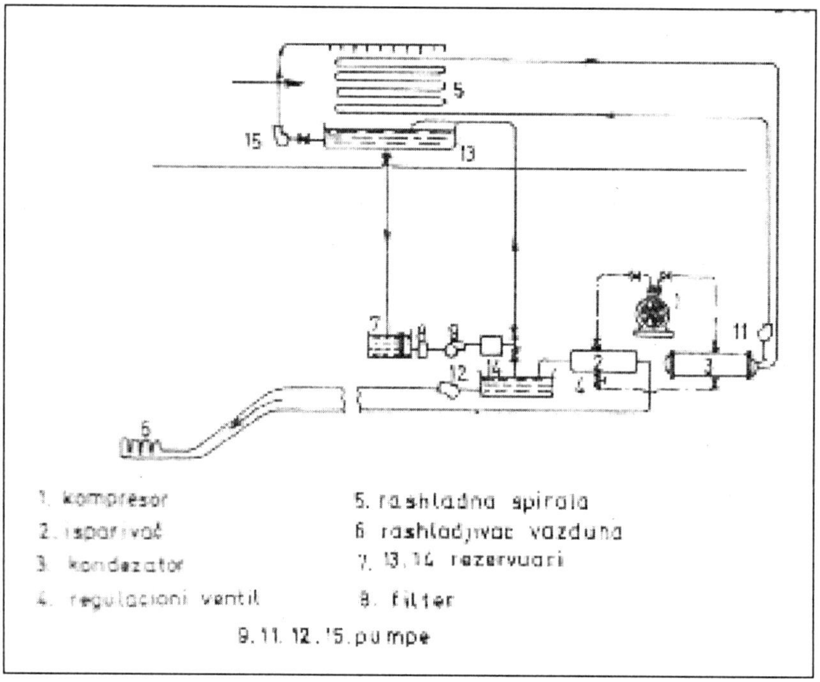

Picture 4

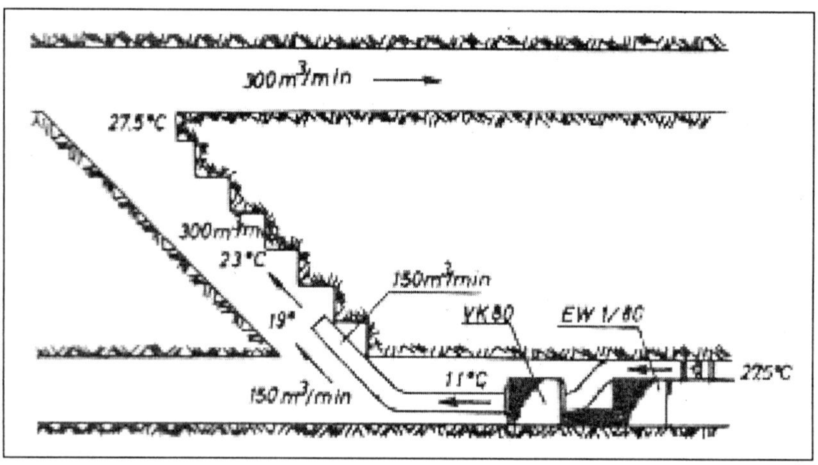

Picture 5

CONCLUSION

With exploitation depth increasing many working conditions is growing worse, among other climatic conditions too. If no ventilation methods can provide better climate conditions in underground mine, air conditioning is needed. The artificial air conditioning can significantly increase expenses and bring further exploitation in question. Air conditioning presents potential problem in future exploitation of the "Borska Reka" ore body that has not been considered enough. It is necessary, above all, to update the data about ore body thermal characteristics, and, after revisioning calculations, if an air conditioning is a must, the consultations with experts, who solved this kind of problems in their mining companies, must be kept before applying any of concrete solutions.

REFERENCES

[1] Miodrag Miljković: Ventilation of a Mine. Bor.
[2] Vesna Jovičić: Ventilation of a Mine. Belgrade 1973.
[3] Tomo Vukobratović: Establishing the Depth of an Underground Mine by Which Air Conditioning Is Not Required. Bor, December 1993.

International Mining Forum 2009, Sobczyk, Kicki & Saługa (eds) © 2009 Taylor & Francis Group, London, UK, ISBN 978-0-415-80428-8

Methodology for the Selection of Optimal Blastholes Length in Underground Mining

Radoje Pantović, Živorad Milićević, Saša Stojadinović
University of Belgrade, Technical Faculty at Bor, Republic of Serbia
e-mail: pan@tf.bor.ac.rs

ABSTRACT: Expansion of high – productive underground mining methods is significantly related to development of drilling equipment. Application of deep blastholes provided reduction of development ratio. On the other hand, deep blastholes are limited by capacity of drilling equipment and worsening of blasted ore fragmentation due to deviation of blastholes.

This paper presents the authors ten year research results of long blastholes application in underground extraction of copper ore in underground mine *Jama Bor.*

In the analysis of *Tilva Roš* and P_2A ore bodies in Jama Bor underground mine, it was shown that increase of oversized ore fragments is proportional to increase of blastholes depth. It means that deviations of deep blastholes would be the most important limiting factor in their future application in *Borska Reka* ore body.

The Paper provides methodology for selection of optimal blastholes depths. Main criterion for selection is a sum of costs, for drilling, blasting, development drifting and secondary blasting. Relation between costs and blastholes depths is shown through: development ratio, drill capacity, consumption of explosives, parameters of blasted ore fragmentation. The result is that increase of drilling accuracy provides increase of optimal blasthole depth.

The significance of this optimisation is in the fact that efficiency of drilling equipment and parameters of designed mining method depend on drilling depth.

KEYWORDS: Underground mining, deep blastholes, drilling capacity, deviation, parameters of blasted ore fragmentation, *Jama Bor*

INTRODUCTION

The application of deep blastholes in underground mines is increasing in past period. The positive economical effects due to the decay of preparation extraction drifts amount justifies the trend of increasing sublevel, level and block heights. On the other hand, problems linked to the long blastholes drilling such as decrease in drilling speed and productivity and blasthole deviation, limit the length of blastholes.

The underground extraction of copper ore in Bor is going on in the "old" part of the mine, above the level K-75 m in the *Brezanik, Tilva Ros* and P_2A ore bodies. The room and pillar mining method with horizontal seam ore extraction is applied in the *Brezanik* ore body. The *Tilva Roš* and P_2A are mined by "Swedish variant" of sublevel caving method. The ore is blasted with 6–8 deep blastholes (Fig. 1). The blastholes are vertical, 76 mm in diameter and their maximal length is 22,5 m. The Simba H 253 drill with COP 1238 hammer is used for drilling.

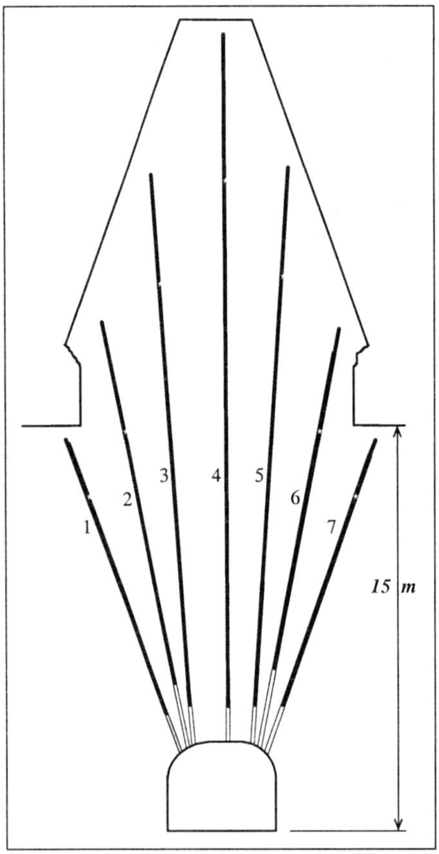

Figure 1. The position of blastholes in *Tilva Roš* and P_2A ore bodies in *Jama Bor* mine

The new part of the mine is in global scale big, ore body *Borska reka*, located below the level of K-78 m. This ore body represents the long term perspective of underground copper ore mining in Bor. The exploration of the ore body and further research showed that precondition for economical extraction of low grade ore from this ore body is the application of high productive mining methods and long blastholes.

During several years of research [1], [2], [3], [4] in copper mine Bor, in *Tilva Roš* and P_2A ore bodies, the parameters of drilling and blasting are recorded, and a part of these records is presented in the paper.

THE INFLUENCE OF BLASTHOLE LENGTH TO TECHNO-ECONOMICAL DRILLING INDICATORS

The choice of optimal blasthole length is of high significance for techno-economical indicators of mining, considering the fact that not only drilling costs are dependant of blasthole length but also the costs of other technological processes, and ore recovery and dilution.

The analysis of blasthole length influence on techno-economical indicators of mining is the base for reliable definition of rational mining method parameters.

Drill productivity

With the increase of the blasthole length, the conditions of impact energy transfer deteriorate, energy loses grow and drilling speed decreases. Empirical formulae for average mechanical drilling speed, from collar to final length, for hydraulic hammers is established upon previously mentioned research results [1]:

$$v_{sr} = \frac{32 \cdot I_{b} \cdot P_u \cdot d^{-1,28} \cdot \left(1 - e^{-0,02 \cdot L}\right)}{L} \quad \text{[m/min]} \tag{1}$$

where: d – blasthole diameter [mm], L – blasthole length [m], I_b – rock drillability index, P_u – impact power of the hammer [kW], δ – drilling speed decrease indicator. Upon the chronometric records of the drilling process in *Jama Bor*, the value of this indicator is set to δ = 0,02.

Figure 2 shows the change in exploitation productivity of the drill in function of blasthole length (N_b – number of blastholes that can be drilled without relocation of the drill). The productivity of the drill rises with the increase of the blasthole length for 10–25 m, due to relative decrease of the auxiliary operations share in total drilling time. However, for larger lengths the productivity decreases due to decrease in mechanical drilling speed. The productivity of the drills is, of course, smaller in rocks with smaller drillability index (I_b).

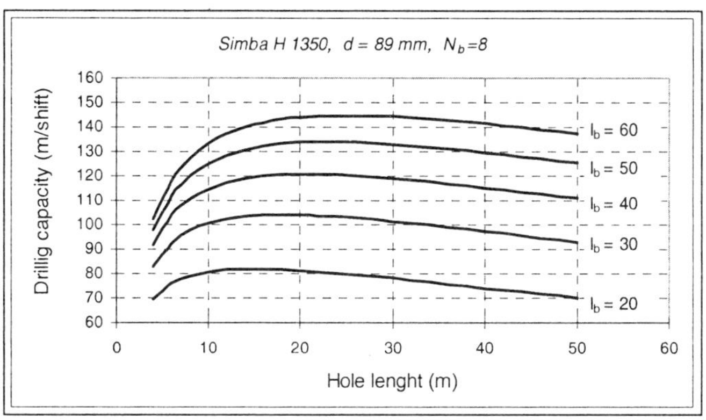

Figure 2. Influence of blasthole length on drilling productivity

Long blastholes deviation

The difference between real and projected position of the blasthole is its deviation or the drilling error. The causes of blasthole deviation can be classified into two groups. The first group is the factors regarding positional and angular errors during the positioning of the drill and boom into the dril ling position and drill bits slipping at drilling start. The second group is the factors leading to the turning of the blasthole direction in the rock mass such as structural heterogeneity of the massif, type of drilling set, pressing force, rotation direction. According to literal sources [6], [8], the initial blasthole deviations are 1–2%, and inside rock mass deviations are up to 5%.

During the process of drilling and loading of blastholes in *Jama Bor* mine the penetrations of blastholes into the caved waste rocks was often registered. The causes of this phenomenon are large blasthole deviations leading to the explosive spreading and loading control problems.

The angular measurements of the horizontal plane and the directions of the visible sections of the blastholes were performed in three extraction drifts (two in *Tilva Roš* and one in P_2A ore body) as follows:

- in planes perpendicular to drifts direction (plane I);
- in planes parallel to the drifts direction and corresponding blastholes (plane II).

Figure 3 shows the designed and real positions of the blastholes in these two planes. The initial angular offset of blasthole direction is the difference between the angles given in drilling order and measured angles.

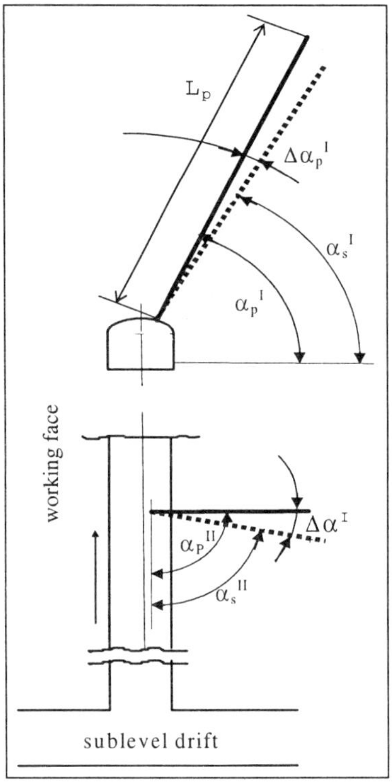

Figure 3. Blasthole deviations: α_p^I and α_p^{II} – designed angles;
α_s^I and α_s^{II} – measured angles and $\Delta\alpha^I$ and $\Delta\alpha^{II}$ – initial angular offsets

The results of the measurements are given in Table 1. The statistical processing of the data gave the parameters of Gaussian distribution, mean value ($\Delta\overline{\alpha}_p$) and the dispersion of initial angular offset ($\sigma^2(\Delta\alpha_p)$).

Table 1. Results of initial blasthole deviations measurements [4]

Ore body	Number of samples	Statistical indicators of initial angular offsets			
		$\Delta\alpha'_p$ (plane I)		$\Delta\alpha''_p$ (plane II)	
		$\overline{\Delta\alpha'_p}$	$\sigma^2(\Delta\alpha'_p)$	$\overline{\Delta\alpha''_p}$	$\sigma^2(\Delta\alpha''_p)$
Tilva Roš – OH 1 Z2, K –16 m	107	0,22	$3{,}00^2$	-0,1	$2{,}02^2$
P_2A – OH 21, K +30 m	54	0,55	$3{,}17^2$	0,5	$2{,}13^2$

By random sample transformation ($\Delta\alpha_p$) Gaussian dispersion can be transformed to normal normalized distribution and the probability that angular offset will fall into specific range can be calculated. The Table 2 gives the probability that absolute deviations of blasthole toes (l_o) will have specific value.

Table 2. Probability of absolute blasthole deviations for the length of 20 m

Ore body	P (l_o < 0,5 m)	P (l_o > 0,5 m)	P (l_o > 1,0 m)	P (l_o > 2 m)
Tilva Roš – OH 1 Z2, K –16 m	0,284	0,716	0,466	0,146
P₂A – OH 21, K +30 m	0,274	0,726	0,484	0,162

The influence of blasthole deviation to the quality of blasting fragmentation

The result of deviation can be intersection, crossing, dispersion or concentration of blastholes in specific area. This leads to the change in burden, disturbance in explosive distribution in the rock mass and deterioration in blasting fragmentation quality.

The influence of blasthole deviation to the blasting fragmentation must be analyzed before the application of high-productive mining methods. This high productivity can not be achieved if boulder content is above 5%.

The boulders would appear even in the case that designed positions of blastholes are achieved and the deviation leads to additional boulder content growth. The influence of blasthole deviation to blasting fragmentation can be expressed trough the ratio of content of boulders due to deviation and general boulder content as:

$$\Delta N = \frac{N_s - N}{N} \cdot 100 = \frac{l_o \cdot \sqrt{W^2 - \frac{l_o^2}{12} - \frac{l_o^2}{6}}}{W^2 - 121 \cdot d^2 \cdot \varepsilon^2 0} \cdot 100 \; [\%] \qquad (2)$$

where: ΔN – boulder content growth due to blasthole deviation [%], W – burden [m], e – explosive relative strength factor, and l_0 – absolute blasthole toe deviation [m], expressed trough angular:

$$l_0 = L \cdot tg(\Delta\alpha) \; [m] \qquad (3)$$

$\Delta\alpha$ – total initial blasthole deviation [°].

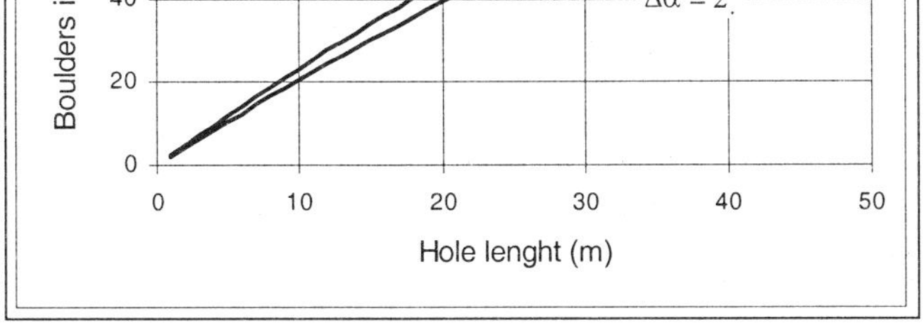

Figure 4. The influence of drilling accuracy to boulder content growth [1]

For the mining conditions in *Jama Bor* mine (d = 76 mm; W = 2 m; e = 1), Figure 4 shows the graph of function (2), with absolute deviation expressed trough angular (Expression (3)). It can be seen from the graph that the increase of the blasthole length would lead to almost linear increase in the boulder content. For example, for 40 m blastholes the boulder contents would be 72,5% and 80,5%, respectively. These considerations show that blasthole deviations are significant from the blasting fragmentation aspect.

Drilling costs

Drilling costs consist of drilling equipment depreciation, labor, energy, drilling set, material, maintenance and servicing.

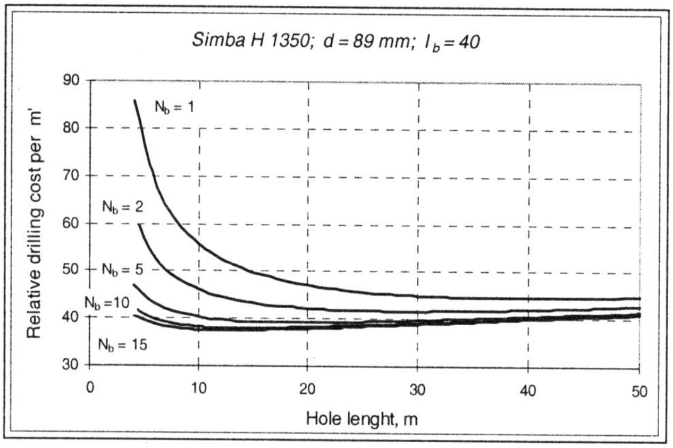

Figure 5. Relation of drilling costs and blasthole length for different drill relocation frequencies

One part of shift drilling costs is fixed (equipment depreciation, maintenance and servicing, labor, fuel, oil and tyres), the other part depends upon drill productivity (drilling set costs) and the third part depends upon mechanical drilling speed (energy costs). Besides that, drilling costs depend on operator training, servicing and maintenance quality, working conditions and other factors.

Drilling costs per ton of ore are dependant on powder factor, diameter and length of blastholes, rocks drillability and other factors since ore extraction factor is a function of drilling and blasting parameters.

Figure 5 shows the relation between drilling costs per one meter of blasthole, for *Jama Bor* mine conditions, and blasthole length, for different number of blastholes that can be drilled without drill relocation. At frequent drill relocation, with parallel blastholes (N_b = 1 and 2), drilling costs are larger, especially for small blasthole lengths. The influence of drill relocation frequency decreases with the increase of the blasthole length.

OPTIMISATION OF LONG BLASTHOLES LENGTH

The sum of drifting (T_p), drilling (T_B), blasting (T_m) and secondary blasting (T_s) costs, represents the criterion for optimal drilling parameters choice, when, upon technical-technological parameters, conditional fragment size is chosen. Mathematical base of the model of blasthole length optimization is the link of the sum of mentioned (T) and the drilling parameters in the following basic shape [5]:

$$T = \varphi\left(d, L, q, N_b\right) = T_p + T_B + T_m + T_s \qquad (4)$$

Cost analysis of single processes indicates that they, directly or indirectly depend on drilling parameters. The influence of blasthole diameter and length, as basic drilling parameters, on total costs sum is both directly and indirectly expressed over powder factor and boulder content.

Drilling costs directly depend on drilling parameters, whilst the blasting costs, over powder factor, indirectly depend on diameter and position of blastholes. Secondary blasting costs are linked to drilling parameters over boulder content, and the costs of drifting depend on blasthole length over sublevel and block heights.

For specific ratio of blasthole diameter and powder factor, percentual contents of boulders can be calculated, for different values of blasthole diameters and lengths. Calculated data, for specific conditions, can be used for determination of the correlation in the shape:

$$N = \varphi_N(d, L) \tag{5}$$

Definition of the costs function is a complex problem which demands comprehensive techno-economical analysis that needs to express the blasthole length influence to the changes in development ratio, drill productivity, fragmentation etc.

The cost function can be partially expressed as a function of blasthole diameter or length, which simplifies the procedure of optimal values definition for single parameter when the other is given. Equalization of partial differential of cost function per blasthole length, as optimized sample, for specific blasthole diameter, results in an expression that can be used to determine the optimal blasthole length (L_{opt}):

$$\frac{\partial T}{\partial L} = 0 \Rightarrow L_{opt} \tag{6}$$

Dependence of total costs from blasthole length, for different drilling accuracy is shown in Figure 6. Due to higher secondary blasting costs, for larger blasthole deviations, the total mining costs are higher. Increase of the drilling accuracy, besides the lowering of mining costs, increase the optimal blasthole length. Dashed line in Figure 6 shows the difference in costs under theoretical assumption that blastholes can be drilled without deviation. In that, ideal case, mining costs decrease even for blasthole lengths higher than 50 m.

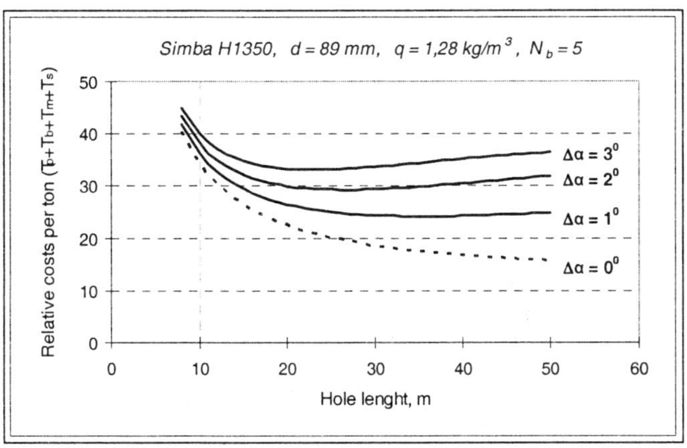

Figure 6. Determination of optimal blasthole length over drilling accuracy

Considering present state of the drilling technique for long blastholes drilling in underground mining, the choice are drills with boom hammers or downhole hammers. Both drill groups have their advantages and lacks but the choice can not be done based only upon their technical characteristics, but on detailed comparative analysis for specific mine conditions.

To illustrate the procedure of drill choice, for given conditions, in Figure 7 two drills have taken into consideration:

- Simba H 4450 with two hydraulic hammers COP 4050, and
- Simba 260 with downhole hammer COP 42.

Average deviations for the first drill are $\Delta\alpha = 2°$ and for the second $\Delta\alpha = 1°$.

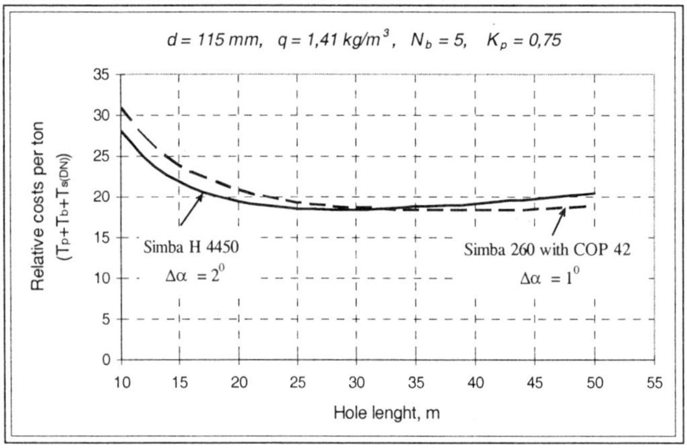

Figure 7. Application area of two drill types

It can be seen from the figure that Simba H 4450 application provides lower costs for blastholes up to 30 m, and for the longer blastholes, the lower costs are achieved with Simba 260 due to hig-her drilling accuracy.

Over the influence on fragmentation, blasthole deviations are basic limiting factor in tendencies to increase blasthole lengths. In addition, blasthole deviations lead to bending and rapid wear of the drilling rods and bits, blasthole walls damage and drill set jams. Higher drilling accuracy develops preconditions for blasthole length increase and resulting cost decrease.

For transfer to high-productive mining methods, which demand application of significantly longer blastholes, needed for ore body *Borska Reka* extraction, it is necessary to provide higher drilling accuracy and apply adequate drilling system.

Determination of optimal drilling parameters for long blastholes other, non costs related factors must be taken into account. Such factors are the risk of face caving and blasting safety [7].

CONCLUSION

The choice of optimal long blasthole length is necessary precondition for successful application and optimization of high-productive mining methods and decrease in mining costs.

The development of drilling equipment widened the possibilities to increase blasthole lengths, increase productivity and decrease mining costs. All this benefited to the development of new, more productive mining methods that resulted in transfer from sublevel to level caving.

The perspective of underground mining of copper ore in the Bor deposit is linked to the application of long blastholes. The problems of drilling and long blasthole application were the issue of several year long researches.

Dependence of some techno-economical long hole drilling indicators (drilling productivity, blasthole deviation and drilling costs) are displayed in function of blasthole lengths, for specific drilling conditions.

Based on the results of measurement of initial blasthole deviations, for the conditions of sublevel caving in *Jama Bor* mine, growth of boulder content due to deviations was calculated.

The criteria for the determination of optimal blasthole lengths were minimal sum of drilling, blasting, drifting and secondary blasting costs, expressed as a function of drilling parameters. Mentioned costs are, directly or indirectly over development ratio, powder factor and fragmentation indicators expressed in the function of blasthole lengths. The result was that optimal blasthole length determination mostly depends on the possibilities of the drilling equipment, considering the drilling accuracy.

Blasthole deviations and resulting decay of fragmentation are the basic limiting factors in blasthole length increase. Higher accuracy drilling can provide better explosive distribution in the massif, better fragmentation and decrease in total mining costs.

REFERENCES

[1] Pantović R.: Determination of Optimal Parameters of the Long Blast Hole Drilling at the High-Productivity Underground Mining Methods. Ph.D disertation, University of Belgrade, Technical Faculty at Bor, 1999.

[2] Pantović R.: Studija smanjenja troškova i uvođenja novih tehnologija u proizvodnju i preradu rude bakra u RTB Bor. Deo: Analiza postojećeg stanja po fazama tehnološkog procesa i definisanje uslova za smanjenje troškova (Tehnologija bušenja i miniranja u pogonu "Jama"), Tehnički fakultet, Bor 2001.

[3] Pantović R.: Studija praćenja, racionalizacije i usavršavanja tehnologije bušenja i miniranja u Jami RBB Bor, Tehnički fakultet, Bor 2002.

[4] Pantović R., Kašić V., Đorđević S.: Rezultati merenja devijacija minskih bušotina na otkopima u jami Bor. XXXIII oktobarsko svetovanje rudara i metalurga, Zbornik radova, str. 114–118, Bor 2001.

[5] Pantović R., Milićević Ž.: Funkcija troškova bušenja dubokih minskih bušotina bušilicama sa hidrauličkim bušaćim čekićima pri podzemnom otkopavanju. Drugi jugoslovenski simpozijum sa međunarodnim učešćem: Bušenje i miniranje, Zbornik radova, str. 68–75, RGF, Beograd 2001.

[6] Horton N.: Straight, No Chaser. WME, No. 7–8, pp. 12–20, 1992.

[7] Nilsson D.: Optimum Stope Height and Blasthole Lenght at Luossavaara. E&MJ, 1986, 6, pp. 52–57.

[8] Valuhin J.K. et al.: Ocenka vlijanija otklonenij skvažin na drobl'enije rudy. Gornyj žurnal, 1979, 3, str. 36–40.

International Mining Forum 2009, Sobczyk, Kicki & Saługa (eds) © 2009 Taylor & Francis Group, London, UK, ISBN 978-0-415-80428-8

Influence of Depth Mine Workings on the Pollution of Mine Water and Possibility Its Limitation

Irena Pluta
Central Mining Institute, Katowice, Poland
e-mail: snxip@gig.katowice.pl

ABSTRACT: Natural mine waters containing the components causing contaminantion of waters in the environment flow into mine workings of the Upper Silesian Coal Basin. Salinity (sodium and chlorides) and barium constitute the most essential, uppermost components, because their concentrations in mine waters exceed, from several hundreds to more than thousand times the threshold pollution values required by the Polish law regulation. Their concentrations increase with lowering the depth of mine workings. The mining methods and technologies applied in mines can be used for limitation of pollutants present in mine water.

KEYWORDS: Mine waters, salinity, barium, radium

1. INTRODUCTION

In the Carboniferous formation of the Upper Silesian Coal Basin (USCB) in Poland mine water with different chemical composition, from fresh water to brines occur. Majority of them contain high concentration of many components [26], [17]. Frequently their amounts belong to higher ones appearing in underground waters in the world. Generally, if chemical composition of mine water complies with the requirements concerning potable water quality these may be used as drinking water (according to the Polish law regulation in Statute Journal No. 61 item 417 of 2007) or for production of spring waters, table waters or mineral waters of natural origin (according to the Polish law regulation in Statute Journal No. 120, item 1256 of 2004). It is also possible to use the mine waters of the USCB as waters of medical usage under the condition that these fulfil the criteria imposed for such waters (according to the Polish law regulation in Statute Journal No. 32, item 220 of 2006). When waters comply with conditions for industrial waters, they may be used in various technological processes and in agriculture. Waters that do not fulfill the requirements set out in the Polish law regulations, as described above, pumped out of the mines to the surface waters (rivers) are classified as industrial waste waters. Some of them are harmful to the water environment. Waters discharged from coal mines in the USCB cause contamination of surface waters: streams, rivers and their sediments in Upper Silesia [1], [3], [14], [15], [17].

Assessing contamination of such mine water should be performed on the basic of comparison of concentrations of pollution indicators with the threshold values for waste waters given in the Order by the Minister of Environment of 24 July 2006 on conditions that have to be fulfilled during discharging waste waters to surface waters or soil and on substances particularly hazardous in (detrimental to) the natural water environment (according to the Polish low regulation in Statute Journal No. 137, item 984 of 2006).

This paper explains influence of the depth of mine workings on the contamination of mine waters and possibility of treatment processes that limit their salinity (majority amounts of sodium and chlorides) and concentration of barium. Particular attention is drawn to the use of mining technologies and methods applied in coal mines of the USCB.

2. GEOLOGICAL AND HYDROGEOLOGICAL CONDITIONS OF THE UPPER SILESIAN COAL BASIN

The Upper Silesian Coal Basin, 7500 km^2 of area (5500 km^2 inside the Poland borders), is situated within the Upper Silesian Variscan depression. The coal-bearing Upper Carboniferous strata includes four litostratigraphic series are 8200 m thick. The series belong to the Namurian A, B, C and Westphalian A, B, C and are composed of sandstones, claystones and coal seams. They are characterized by a gradual reduction of their thickness toward the east and south-east.

The coal-bearing formation covered by Triassic, the Jurassic and the Quaternary deposits in the north-eastern part and by the Miocene, the Triassic and the Quaternary deposits in the southern and western parts. In these regions different hydrogeological conditions occur. Their boundaries are delineated by the extent of the practically impermeable sediments of the Miocene formation. The aquifers are recharged in zones of outcrops or through permeable covering rocks in the north-eastern part of the USCB. Locally recharge takes place also in the central part of the basin, in the areas where the Miocene sediments have been eroded or their thickness is low. The recharge of the Carboniferous water-bearing sandstones by water of the Quaternary from recent and buried valleys is the most active.

The Upper Carboniferous formation is represented by a clay-silt-sandstone complex containing coal seams. The Carboniferous water-bearing sandstones and conglomerates have thickness ranging from 0.5 to several dozen meters. They are isolated one from another by intercalation of practically impermeable claystones, except fault zones, zones of sedimentary wedging as well as areas of mining. in these conditions the decreasing of hydrogeological properties of water-bearing rocks with the increasing depth [26].

3. MINE WATER IN THE UPPER SILESIAN COAL BASIN

The Upper Silesian Coal Basin is located in the southern Poland. The coal has been mined since the second half of the 18th century. Sixty-six hard coal mines were working in this region in the past century. At present, only half of them are still active mines. All mine working is carried out underground. Due to the great number of mineable coal mines in the sequence, the mining operations are carried out simultaneously at several depth levels, covering almost entire surface of mining area. The depth of mine workings varies from 220 to 1160 m.

The differences in the geological structural, hydrogeological conditions, underground coal mining activities of the USCB and origin of water-bearing flowing into mine workings has resulted in a wide range of their salinity and chemical composition mine waters.

The waters flowing into the mine workings constitute a special group of groundwaters of the Upper Silesian Coal Basin. These are natural mine waters that constitute the so-called natural inflow of the underground mine workings. Their chemical composition is formed by natural hydrogeochemical processes taking place in the hydrogeological system of the USCB and may be the effect of the mining activities performed (anthropopression). The influence of anthropopression may be direct or intermediate, and water components which can be of geogenic or antropogenic character.

A detailed analysis of chemical composition of natural mine waters showed that many pollution indicators determined for waste waters (according to the Polish law regulation in Statute Journal

No. 137, item 984 of 2006) concern components that do not exist in the mine waters. Thus it was proposed to simplify the procedure of assessing quality of mine water classified as industrial wastes which discharged to the surface waters and to soil by taking into account only those contamination indicators that are characteristic of mine waters. Based on the analysis of chemical composition of mine water in the USCB thirty physicochemical and chemical components: temperature, pH reaction, suspended solids, total organic carbon, volatile phenols, surface-active substances – anion ones, substances extractable with the use of petroleum benzine, petroleum substances, volatile aromatic hydrocarbons (VAH), ammonia nitrogen, nitrate nitrogen (III, V), chlorides, sulphates, sodium, potassium, general iron, general phosphorus; harmful components: arsenic, barium, boron, general chrome, zinc, copper, nickel, lead, fluorides and mercury, cadmium, free cyanides comparing. So, the degree of contamination may be assessed based on the analysis of their concentrations and by referring these contents to the threshold pollution values set out in the Order by the Minister of Environment (Statute Journal No. 137, item 984 of 2006).

4. POLLUTANTS IN MINE WATER OF THE UPPER SILESIAN COAL BASIN FLOWING INTO MINE WORKINGS ON THE DEPTH UP TO 1000 M

In order to distinguish components causing the contamination of mine water flowing into mine workings of the coal mines in the USCB on the depth up to 1000 m, analyses of two hundred and fifty waters flowing into mine workings were carried out. In all waters physico-chemical and chemical components were determined, whereas in selected waters contamination was analysed as well [17].

Among the components of mine water three physico-chemical pollution indicators appear: temperature, pH and general suspension. In natural mine waters, which were taken in mine workings situated on extraction levels up to the depth of 1000 m, temperature exceeding 35°C has not been ascertained. This value has been determined as the threshold pollution value in relation to mine waters as the waste waters deriving from mine drainage. Similar investigations into total suspensions have not pointed out higher contents than the threshold value, i.e. 35 mg/dm^3. The values of pH in mine waters ranged from 4.6 to 8.9. The presence of waters with pH less than 6.5 can lead to possibility of occurrence slightly acid waters discharged from mines (6.5–9). The above facts indicate that pH is the only physico-chemical pollutant, the values of which can be exceeded in waters flowing into mines.

The pollution of mine water is caused by the following groups of organic compounds: total organic carbon, volatile phenols, surface-active substances (anionic), substances extracted by petroleum benzine, petroleum derivatives, aromatic hydrocarbons (benzene, toluene, xylene). Among them the most important is their total content. It can be defined on the basis of the determination of total organic carbon (TOC). Investigations of this pollution indicator in several waters flowing into mines indicated a maximum content reaching 5.85 mg/dm^3, however, in mine waters deriving from mines of the south-west part of the USCB the concentrations have not exceeded 4.9 mg/dm^3. The determination of other organic compounds, such as: volatile aromatic hydrocarbons, i.e. benzene, toluene, xylene (BTX), volatile phenols as well as anionic surface-active substances (detergents), petroleum derivatives and organic substances extracted by petroleum benzine has indicated that their contents do not exceed the threshold pollution values in relation to mine water as the waste waters in the Polish law regulation (Statute Journal No. 137, item 984 of 2006). Investigations of few natural mine water of the south-west part of the USCB show that benzene concentration reaches 0.01 mg/dm^3, toluene up to 0.018 mg/dm^3 and xylene up to 0.0007 mg/dm^3. These contents are considerably lower than the threshold pollution value determined generally for their sum in mine waters. Similarly as the content determinations of sixteen components belonging to the group of polycyclic aromatic hydrocarbons (PAH) carried out in mine waters by Caban-Pabian [2]. Thus, generally organic compounds occur in low concentrations and they do not cause the contamination of natural mine waters flowing into the USCB coal mines.

Nitrogen compounds are very important pollution indicators. When assessing their impact on the water environment we take into consideration the reduced forms – ammonia (NH_3) and ammonium ion (NH_4^+) as well as the oxidised forms: nitrate ion (V) (NO_3) and nitrate ion (III) (NO_2^-). In the indicator formulation these are: ammonia nitrogen (N-NH_4), nitrate nitrogen (N-NO_3) and nitrite nitrogen (N-NO_2). The determinations of different forms of reduced nitrogen have pointed out that in natural mine waters flowing into coal mines of the USCB the ammonium ion is the main component with concentrations reaching up to 96 mg/dm^3 [12], [26]. These contents exceed several times the threshold pollution value for ammonia nitrogen (10 mg/dm^3).

The determinations of the nitrate ion (V) (NO_3), however, have indicated contents not exceeding several milligrams, and of the nitrate ion (III) (NO_2) one milligram in 1 dm^3 of natural waters flowing into mines. Frequently their concentrations in salty mine water are lower them the determination ability boundaries, i.e. 0.1 mg NO_3^-/dm^3 and 0.02 mg NO_2^-/dm^3. Nitrate nitrogen (nitrate nitrogen (V) N-NO_3) and nitrite nitrogen (nitrate nitrogen (III) N-NO_2) are not pollutants of natural mine waters.

Determinations of the remaining chemical components show that in natural mine waters high concentrations of sodium – up to 68 970 mg/dm^3, potassium – up to 840 mg/dm^3, chlorides – up to 125 470 mg/dm^3 as well as iron ions (II, III) – up to 49.5 mg/dm^3 and sulphates (VI) – up to 8170 mg/dm^3 [10,13,26] occur. These contents exceed their threshold pollution values determined for waters deriving from coal mines in the Polish law regulation (Statute Journal No. 137, item 984 of 2006). Especially, the concentrations of sodium and chlorides are from several dozen up to hundred times more than the threshold values (respectively 800 and 1000 mg/dm^3).

In mine waters of the UCSB metals defined as heavy metals (i.e. arsenic, zinc, chromium, copper, nickel, lead, cobalt, mercury, barium) are present. Among them two metals as particularly harmful for human health and the water environment mercury and cadmium are to be considered carefully. Kulawik and others [7] have ascertained in mines of the Rybnik region lead, copper, and cadmium concentrations not exceeding 0.01, 0.04, and 0.005 mg/dm^3 respectively. Generally the concentrations of arsenic, total chromium, zinc, cadmium, copper, nickel, lead, and mercury in few tested natural mine waters have not exceeded the threshold pollution values for waste waters discharged from mines. Among heavy metals in natural waters flowing into mine workings of the USCB, in considerable quantities, exceeding their threshold pollution value, only barium was ascertained [10], [19]. The determination of this component has pointed out that its concentration reaches 2400 mg/ /dm^3. This value is above thousand times more than the threshold pollution value in mine waters according in the Statute Journal No. 137, item 984 of 2006 (2 mg/dm^3). Thus barium causing the contamination of natural mine waters and its content should be reduced.

Besides barium, a special pollution indicator of mine water is boron. In waters flowing into coal mines of the southern part of USCB the content of this pollutant reaches 7.5 mg/dm^3 [17]. According to Różkowski [26] in natural mine waters of the Carboniferous formation boron concentration can reach up to 30 mg/dm^3, and in the Miocene waters according to Jarocka [5] up to 27.2 mg/dm^3. The concentration of boron in mine water is several times more that threshold pollution value in waste waters (1 mg/dm^3).

The investigations regarding fluorides in few waters flowing into mines have indicated contents not exceeding some mg/dm^3. Generally, they are contained in the range up to 9.3 mg/dm^3, which for this component Pałys [9] has ascertained in mine waters. Similarly the analyses of total phosphorus, determined in the form of phosphates, have pointed out its low contents amounting to several milligrams in 1 dm^3 of waters flowing into mines. Generally the contents of fluorides and phosphorus in mine waters of the USCB do not exceed the threshold values of these pollution indicators for waters pumped from coal mines.

In underground waters nitrogen, besides the described above ammonium and nitrate (III, V) forms can be present in the form of cyanides. Cyanides occur as free ions CN^- or bounded ions in complex compounds. On account of the harmfulness, mainly the contents of free cyanides in mine waters of the USCB were determined. Investigations of this component in salty mine waters diluted to the content of 1 g/dm^3 of chloride ion practically have not indicated their presence.

As mentioned above, mine waters of the USCB flowing into mine workings up to depth 1000 m contain components, which cause the contamination of surface waters. Contents of potassium, ammonium nitrogen, sulphates, boron, iron and pH are few times, but in the cases of sodium, chlorides and barium, from several dozen to above thousand times more than the threshold pollution values defined for mine waters discharging from coal mines, in according to the Order by the Minister of Environment (Statute Journal No. 137, item 984 of 2006).

5. POLLUTANTS IN MINE WATER OF THE UPPER SILESIAN COAL BASIN FLOWING INTO MINE WORKINGS ON THE DEPTH BELOW 1000 M

According to the investigation of the pollution indicators in mine water flowing into mine workings of the coal mines in the USCB on the depth up to 1000 m, which presented above, analyses of few mine waters from the depth below 1000 m were performed. In the Table 1 the maximum concentration of essential pollutants i.e. sodium, chlorides and barium causing the uppermost contamination of mine waters flowing into coal mines of the USCB are presented. These values are higher than in natural mine waters flowing into mine workings on the depth up to 1000 m. The highest barium content was ascertained in water flowing into the Zabrze coal mine, but sodium and chlorides in the Marcel coal mine.

In these waters concentrations of zinc, lead, and copper reaches up to 5.5, 1.5 and 0.34 mg/dm^3 respectively were ascertained. These values are similar to those pointed out by Wagner [27] in the waters from the coal-bearing formations in 190 exploration boreholes that were drilled from 1953 till 1990 and in the waters from boreholes: Wygorzele, Czarków, Rudołtowice, Mizerów, Urbanowice drilled from 1995 till 1998.

Table 1. Maximum concentration of sodium, chlorides, barium and heavy metals (Cd, Cr, Cu, Pb, Ni, Hg, Zn) in waters of the USCB from the deepest levels, below 1000 m

Components [mg/dm^3]	In waters from boreholes drilled 1953–1990 according Wagner [27]	In waters from boreholes drilled 1995–1998	In mine waters flowing into mine workings	The threshold pollution value for waste waters according the Order of 24 July 2006
Sodium	n.a.	57770	96100	800
Chlorides	138000	113725	203150	1000
Barium	n.a.	2400	2870	2.0
Cadmium	0.75	0.1	< 0.01	0.4
Chromium total	0.15	< 0.1	< 0.03	0.5
Copper	0.25	0.2	0.34	0.5
Lead	1.8	0.9	1.5	0.5
Mercury	n.a.	n.a.	< 0.03	0.2
Nickel	1.4	0.2	0.07	0.5
Zinc	4.6	2.0	5.5	2.0

n.a. – not analyzed

6. TREATMENT MINE WATERS WITH SALINITY (MAJORITY OF SODIUM AND CHLORIDES)

In the 1950s, when it was ascertained that salty waters and brines from the mines cause the pollution of surface waters (rivers) desalting and the hydrotechnical method as preventive means salinity impact (limiting the sodium and chlorides), were proposed. Mine waters desalting leading to cause drinking or industrial water and sodium chloride were used in the Dębieńsko and Budryk coal mines [8]. The hydrotechnical method enabling controlled discharge of mine waters into surface waters was applied in the south-western part of the USCB system means Olza collector and in the eastern part in the Janina, Brzeszcze, and Silesia coal mines. In the 1990s in order to reduce the impact of mine waters, their injection back into the Upper Carboniferous deposits was proposed (into formations occurring at a great depth), the so-called underground injection or recirculation (into water-bearing layer drained by mine workings), as well as brine injection from mine workings to porous and permeable layers. These operating modes were tried in the coal mines: Krupiński, Czeczott, Piast and Jaworzno [24], [25].

The methods described above enabled the limitation of salinity into surface waters which flow in the Upper Silesia, but did not solve this problem in full. Therefore in connection with the above in 1980s the use of mining methods and technologies in mine workings was proposed. They consist of the modification of mining operations conducting in order to cause the reduction of inflows, especially of the brines. This can be obtained among others by the change in mining works (excluding from operation the mine workings into the brines flow). For the limitation of inflows the closing of water and methane drainage boreholes, damping of mine workings out of extraction, sealing of mine workings with intensive inflowing of brines or storage of waters in cavities caused by the extraction can be also used. Unfortunately the possibility of these methods depends on local mining and hydrogeological conditions; therefore their application is not general.

In the 1990s in coal mines of the USCB water suspensions of fine-grained industrial wastes (fly-ashes, products from desulphurisation processes, flotation tailings) have been used in mine workings as fire- and methane-protection agents. These wastes filled of cavities, gobs and underground old mine workings by the wastes from power plants and mines. These cavities cause the sealing of mine workings with respect to the inflow of natural mine waters. As the effect, in the coal mines of the Jastrzębska Coal Company (JSW), where the use of the flotation tailings as well as wastes from power plants in 1991–2000 was common limitation of about 20% of chlorides (as well as sodium) in mine waters discharged into surface waters took place (the Oder river) [21].

In the 2000s was obtained (it was proposed) that the highly saline waters from active coal mine should be disposed the old mine workings, gobs and cavities into the neighbouring, closed coal mine. This situation can be realized in the south-western part of the USCB. Brines flowing into mine workings of the Marcel coal mine are directed into to the abandoned 1-Maja coal mine, which is connected with the Marcel coal mine [22].

Actually, the best results have been too obtained by the hydrotechnical method enabling controlled discharge of mine water into Odra and Wisła rivers in the south-western and eastern part of the USCB [23] and the disposal of brines from the Marcel coal mine into the abandoned 1-Maja coal mine.

7. PROPOSED TREATMENT OF MINE WATER IN COAL MINES OF THE USCB

It is preferred to carry out the treatment mine water "at source" (reduction "at source"). The component causing mine water contamination should be reduced or removed in underground mine workings. The treatment of mine water on the surface, for instance in reservoirs, settling ponds is possible, but this process requires special control, waste disposal or costly management. Therefore, the purification of mine waters in reservoirs or settling ponds is not recommended.

The special chemical or biological methods of treatment of mine water is difficult because the mine workings are of small sizes and expensive. In connection with the above most advantageous is the use of processes, which support the self-purification of mine waters, i.e. physical, chemical or biological processes, taking place during their flow in mine workings. The best method is the initiation of these processes in places where natural mine waters get into mine workings, thus the removal of pollution components *in situ* in the water-bearing layer or limitation of their amount during flow through a "protective barrier", "purifying barrier". Attention has been drawn to the possibility of purifying mine water by means of the mining methods and technologies applied in coal extraction.

8. PURIFICATION OF MINE WATER OF THE USCB WITH BARIUM

Among components causing the pollution of mine water in the USCB, barium is especially harmful. Its concentration in waters flowing into mines exceeds more than one thousand times the threshold pollution value, determined for waste waters pumping from mines in Poland [17], [19]. Its content in mine waters causes technical problems, consisting in the pollution ("overgrown" of pumps, pipe, and pipelines) of water transport systems. The simplest, and simultaneously the most effective method of removal of barium from min water is its precipitation in the form of practically insoluble deposits, mainly sulphate.

Thus the factors purifying the natural mine waters are first of all sulphates which reduce the amount of barium. Their source in mines can constitute different sulphur compounds contained in the Upper Carboniferous strata or in covering deposits, as well as substances used in mining operations. The most important process of sulphates generation is the oxidation of sulphides. As the effect of dissolving, leaching and oxidation, sulphates get into mine waters.

In the Upper Silesian Coal Basin sea sediments of the Miocene and the Triassic occur. Some of them contain sulphate minerals of calcium: gypsum ($CaSO_4{\cdot}2H_2O$) and anhydrite ($CaSO_4$). Their dissolving enriches the mine waters with sulphates.

In mine waters can also dissolve sulphate minerals contained in Upper Carboniferous formation. Gypsum, anhydrite in different lithostratigraphic series occur [6]. Besides them in the Upper Carboniferous coal-bearing deposits also hydrated sulphate minerals: epsomite ($MgSO_4{\cdot}2H_2O$), jarosite ($(Na, K)Fe_3(SO_4)_2(OH)_6$), melanterite ($FeSO_4{\cdot}7H_2O$) in the form of crusts and powder aggregates, mainly in fracture zones exist.

In the Upper Carboniferous deposits of the disulfides as pyrite and marcasite (FeS_2) occur. Under certain conditions these substances can become source of sulphates for the mine waters in contact with the coal-bearing strata. Sometimes this process is observed in water that flow through the cavities caused by the extraction, gobs, and old mine workings. This process was noticed in the USCB for the first time in the Silesia coal mine [20], [16]. The results indicate that the desulphurisation process of the Carboniferous formation reduce the amount of barium which is content in natural mine waters in the Silesia coal mine.

Sulphates deriving into mine waters among others from sulphate minerals (products) – gypsum and anhydrite. These substances are have been used in some mining technologies. Gypsum and anhydrite are used in coal mines of a sealing mixture for fire protection. Anhydrite is also used for the protection of the mine working places against overpressure of the Carboniferous rocks. These technologies were used in the Morcinek and Jastrzębie coal mines. Sulphates dissolved into natural mine waters and precipitate the barium. The treatment process with barium depends on amount of gypsum or anhydrite used and duration of contact with the fire-protection and overpressure walls.

As the described above, from the half of the 1990s, the fine-grained industrial wastes in coal mines have been utilized. They have been used in different mining technologies, among others for self-consolidating filling, for sealing of longwalls and roadways, for performing of plugs and dams as well as roadside packs. Wastes from power plants constitute mainly fly ashes and flue gas desulphurisation products. These wastes contain sulphates: anhydrite ($CaSO_4$) and gypsum ($CaSO_4 \cdot 2H_2O$) as well as hannabachite ($CaSO_3 \cdot 1/2H_2O$) [4]. From these wastes in mine workings sulphates get into mine waters [11]. Sulphates in the mine waters draining off from storage of wastes have shown that their concentrations can reach 4150 mg/dm^3. Their amounts depends on the mining technologies, the type of wastes, the type of coal used for combustion and the hydrogeological conditions in the mine workings.

The post-extraction cavities, gobs and old mine workings are filled with mining wastes. The wastes constitute Carboniferous barren rocks, separated from the extracted in coal cleaning preparation. In these processes the fine-grained wastes derive from flotation and washing processes, as well as sediments from drainage roads. In these wastes similarly as in wastes from power plants sulphates occur. They derive from oxidation processes of sulphur minerals contained in the Upper Carboniferous formation. Generally their amounts are lower than in the case of the described above wastes originating in power plants. For example, into 1 dm^3 of mine water in the Borynia coal mine about 500 mg of sulphates release, from the flotation tailings.

Sulphates contained in waters leaching from the places of power and mining wastes using in different technologies cause barium precipitation from natural mine waters flowing into mine workings. These favourable effects in the reduction of the barium content in mine water observed during the storage of fly ashes in the Pniówek coal mine [18].

SUMMARY AND CONCLUSIONS

Into mine workings of the Upper Silesian Coal Basin, and to the surface waters (rivers) of the Upper Silesia in Poland flow natural mine water containing components, some of them cause contamination of waters in the environment. Salinity (majority of sodium and chlorides) and barium constitute the most essential, uppermost components, because their concentrations in mine waters exceed from several hundreds more to thousand times the threshold pollution values required by the Polish law regulation (Order of Minister of Environment 2006 in Statute Journal No. 137, item 984 of 2006). Their concentrations increase with lowering the depth of mine workings. Therefore their contents should be reduced.

The mining methods and technologies applied in mines can be used for limitation of pollutants present in mine water. They are an example of the most favourable reduction of pollution "at the source" of its arising, i.e. underground, and they do not need any capital outlays.

Use of mining methods and technologies in purification of mine waters this is an example of connection in ecological and mining interests.

In some mine waters flowing into mine workings on the levels below 1000 m the concentration of pollutants, heavy metals: lead and zinc are higher than their threshold pollution values.

REFERENCES

[1] Bojakowska J., Sokołowska G. 1998: Influence of Mining and Metallurgy on Pollutants in Sediments of Oder River. (In Polish). Przegląd Geologiczny, No 7, pp. 603–608.

[2] Caban-Pabian B. 2002: Investigations of Selfpurification of Mine Waters from Polycyclic Aromatic Hydrocarbons. (In Polish). Archiwum Ochrony Srodowiska, Vol. 28, No 4, pp. 57–66.

[3] Czaja S., Jankowski A.T. 1992: The Contribution of Mine Waters to the Discharge and Salinity of Rivers in Katowice Province. Archiwum Ochrony Środowiska, No 2, pp. 181–194.

[4] Garbacik A., Spyrka J. 2000: Application Wastes from Flue Gas Desuphurization Processes in Cement Production. (In Polish). Cement, Wapno, Beton, No 1, pp. 8–14.

[5] Jarocka A. 1976: Physicochemical Analyses of Medical Waters, Therapeutic Mud. (In Polish). Centralny Ośrodek Informacji Uzdrowiskowej, Warszawa 1976.

[6] Kuhl J. 1955: Petrografic Classification of Rocks in Carboniferous Formations of the Upper Silesian Coal Basin. (In Polish). Prace GIG, Seria A, Komunikat No. 171, p. 32.
[7] Kulawik M., Szreder M., Pluta I. 1998: Lead, Copper and Cadmium in Waters from the Coal Mines of the South-Western Part of the Upper Silesian Coal Basin on Example Jankowice Coal Mine. (In Polish). Materiały III Konferencji Naukowo-Technicznej pt.: "Ochrona środowiska w aspekcie gospodarki wodnej oraz gospodarczego wykorzystania odpadów i metanu w Rybnickim Okręgu Węglowym", Wyd. Wiadomości Górnicze, Katowice, p. 61–64.
[8] Motyka R., Magdziorz A. 1994: Directions and Progress in Desalination. (In Polish). Wiadomości Górnicze, No. 9, pp. 331–333.
[9] Pałys J. 1969: Possibility Practically Using Microcomponents from Mine Water of the Upper Silesia. (In Polish). Przegląd Górniczy, No. 6, pp. 304–308.
[10] Pluta I. 1988: Radioactivity Waters of the Carboniferous Formation in the USCB and Their Origin from the Isotope Data. (In Polish). PhD thesis, AGH, Kraków.
[11] Pluta I. 1997: Ashes as Purification Agent of Waters in Coal Mines of South, Western Part Silesian Coal Basin. (In Polish). In: "Współczesne problemy hydrogeologii", Vol. VIII, pp. 247–250.
[12] Pluta I. 1998: Ammonium Ion in Waters from Coal Mines in the South-Western Part of the Upper Silesian Coal Basin. (In Polish). Materiały III Konferencji Naukowo-Technicznej pt.: "Ochrona środowiska w aspekcie gospodarki wodnej oraz gospodarczego wykorzystania odpadów i metanu w Rybnickim Okręgu Węglowym", Wyd. Wiadomości Górnicze, Katowice, pp. 71–76.
[13] Pluta I. 2000: Changes of the Chemical Composition in Mine Workings Coal Mines of the USCB. (In Polish). Zeszyty naukowe, Seria Górnictwo; Konferencja nt.: "Rozpoznanie złóż. Górnictwo zrównoważonego rozwoju", H. 246, pp. 407–416.
[14] Pluta I. 2001a: Barium Discharged from Coal Mines and Its Impact on the Environmental of the Upper Silesia (Poland). In: Proceedings Secotox World Congress: Ecotoxicology and Environmental Safety on the Verge of Third Millennium, Trends, Threats and Challenges, Cracow 2001, pp. 41–45.
[15] Pluta I. 2001b: Barium and Radium Discharged from Coal Mine in the Upper Silesia, Poland. Environmental Geology, Vol. 40, No 3, pp. 345–348.
[16] Pluta I. 2004: Acid Mine Waters in Coal Mines of the South-Western Part of the Upper Silesian Coal Basin. (In Polish). Przegląd Górniczy, No. 2, pp. 20–23.
[17] Pluta I. 2005: Mine Waters of the Upper Silesian Coal Basin – Origin, Pollutants and Purification. (In Polish). Prace GIG, No. 865, p. 169.
[18] Pluta I., Waleczek E. 1996: Purification of Brines in the Pniówek Coal Mine from Barium and Radium. (In Polish). In: Materiały VI Konferencji pt.: "Problemy Geologii w Ekologii i Górnictwie Podziemnym", Ustroń, Wyd. GIG, pp. 309–318.
[19] Pluta I., Pałys J. 1999: Barium in Mine Waters of the Upper Silesian Coal Basin. (In Polish). Przegląd Górniczy, No. 4, pp. 16–22.
[20] Pluta I., Zuber A., Pytlak M. 1994: Possibility of Using Old Mine Workings in Purification of Mine Waters. (In Polish). In: Materiały V Konferencja "Problemy geologii w ekologii i górnictwie podziemnym", Szczyrk, Wyd. GIG, pp. 347–353.
[21] Pluta I., Mazurkiewicz M., Łukaszczyk Z. 2002: Utilization of Wastes in Coal Mines Using in Water Protection. (In Polish). Miesięcznik WUG, Bezpieczeństwo Pracy i Ochrona Środowiska w Górnictwie, No. 10, pp. 28–32.
[22] Pluta I., Marcol A., Dziendziel F., Karwasiecka M. 2008: Utilization and Purification of the Mine Waters at the Marcel and 1-Maja Coal Mines in the Upper Silesian Coal Basin (Poland). Mine.
[23] Pustelnik J., Mrowiec G., Sokół A. 2008: Dosing Mine Water in the Collector Olza. (In Polish). Materiały XV Międzynarodowej Konferencji Naukowo-Technicznej pt.: "Ochrona środowiska w planach zagospodarowania przestrzennego miast i gmin w subregionie zachodnim województwa śląskiego", Rybnik 2008, pp. 281–286.
[24] Rogoż M., Posyłek E. 2000: Hydrogeological Problems in Polish Coal Mines. (In Polish). Wyd. GIG, p. 402.
[25] Rogoż M. 2002: Hydrogeology of Mines and Basis of General Hydrogeology. (In Polish). Wyd. GIG, p. 683.
[26] Różkowski et al. 2004: Hydrogeochemical Environment in the Upper Silesian Coal Basin. (In Polish), Wyd. Usl.
[27] Wagner J. 1998: Hydrogeological Characteristic of the Carboniferous Formation General Syncline in the Upper Silesian Coal Basin. (In Polish). Biuletyn Instytutu Geologicznego, No. 383, p. 55–96.

International Mining Forum 2009, Sobczyk, Kicki & Saługa (eds) © 2009 Taylor & Francis Group, London, UK, ISBN 978-0-415-80428-8

The Place and Historic Importance of the "Glogow Gleboki – Przemyslowy" Copper Orebody in the Strategy of KGHM "Polska Miedz" Company

Piotr Saluga
MEERI of the Polish Academy of Sciences, Krakow, Poland

Jerzy Kicki
AGH – University of Science & Technology,
MEERI of the Polish Academy of Sciences, Krakow, Poland

ABSTRACT: The "Glogow Gleboki – Przemyslowy" deep orebody was explored in the years 1962–1986. It is today considered as a basis for KGHM "Polska Miedz" Co. in the coming years, enabling the company to continue production at the current rates. Developing the deposit will also determine the Lower-Silesian Lubin region's existence in the near future. The paper presents a place and a strategic role of the deposit in the company's plans throughout the past decades.

KEYWORDS: KGHM "Polska Miedz", Polish copper, deep deposit, "Glogow Gleboki"

BACKGROUND

The "Glogow Gleboki – Przemyslowy" (GGP) orebody occurs at the depth of 1,100–1,400 metres. It is a deep part of huge copper deposit located in Lower Silesia, Poland, between towns Legnica and Glogow. The orebody conditions, which can be expected when mining in the area, are to a large extent determined by parameters inherent to large depths, e.g. high geothermic gradient. It currently accounts for approximately 26% of the total mineral reserves at KGHM "Polska Miedz" Co. (KGHM "Polish Copper" Co.).

The former mining concession areas assured the mines belonging to KGHM would exist for at least 30 years. However, should they fail to develop new deposits, as soon as in 2015–2020 the mines would see their production drop by as much as 50%.

The management is in no doubt that in order to maintain the current production levels the Company needs to start mining new deposits – either in Poland or abroad.

The decision to develop the GGP reserves was not easy to make. The arguments in favour were, among others, closeness and geological continuity with the deposit currently mined, high metal prices at the international markets and the problems KGHM encountered at its first oversees project, in Congo DR. The arguments against, on the other hand, were difficult mining conditions expected in the area due to high rock temperature and high rockmass stress, as well as unsatisfactory geological knowledge of the deposit.

Achieving full planned production from the GGP tenement will coincide with decline in production in currently mined and gradually exhausted areas. The decision to develop the GGP deposit will allow extending the life of KGHM by additional twenty-odd years. This is of great importance to the Company and the whole region.

KGHM, the biggest Polish copper producer, went a long way from the time the deposits were discovered in the 1950s and the Copper Mining Region established. From 1962, when first exploratory boreholes were drilled into the deep deposit, until 2006, when the decision to develop and mine the deposit was made, 44 years have passed. Hence, the most important ally of the vision to develop these mineral resources turned out to be ...time.

1. AN OUTLINE OF THE HISTORY OF THE DEPOSIT UNTIL THE MID-1990s

In 1957 the biggest copper deposit in Europe was discovered near Lubin and Polkowice in Lower Silesian Region. The discovery came as a result of a wide-scale exploration project covering the area north of the Old Copper Basin deposits and North of Wroclaw. The success would not have been possible without the faith, relentlessness, endurance and commitment of Jan Wyżykowski, the manager of the drilling programme, who was later dubbed the "father" of the discovery (Rydzewski 1996).

The "Lubin-Sieroszowice" deposit was discovered on March 23, 1957, by drilling through a copper bearing Permian orebody containing chalcocite mineralizations in "Sieroszowice IG1" (S1) borehole, and confirmed on August 8, 1957, by "Lubin IG1" (S19) borehole. Subsequent drilling further confirmed the continuity of the mineralization (Rydzewski 1996).

On December 28, 1959, the Minister of Heavy Industry called into being "Lubin Mine Under Construction State Company". It was subsequently renamed on April 5, 1961, to "Mining & Metallurgical Copper Plant under construction" (in Polish: "Kombinat Gorniczo-Hutniczy Miedzi w budowie") and on December 30, 1968, to "Mining & Metallurgical Copper Plant" (in Polish: "Kombinat Gorniczo-Hutniczy Miedzi") (Piestrzynski (ed.), 1996). KGHM "Polska Miedz" Co. is a direct descendant of the company.

The drilling project continuing through the years 1962–1986 served to explore the deep deposit. The exploratory works covering the areas of "Bytom Odrzanski", "Gaworzyce", "Glogow", "Retkow" and "Radwanice-Zachod" confirmed continuity of the New Basin in the North-Easterly direction down to depths of 1,100–2,000 metres. At the time, nobody considered resources of these areas for mining; it was assumed, however, that in due course, as shallower ores were gradually exhausted, technological advancements would allow their extraction.

In the years 1990–1996 three geological reports were prepared for "Bytom Odrzanski", "Glogow" and "Retkow" tenements, their mineral resources classified as C_1 and C_2 categories of the Polish resources & reserves classification system (max estimation error: ±30–40%). The reports did not contain any new information and in this sense they are secondary. They served merely to organize and tidy up the existing data (Banaszak, Tomanik 2002).

2. THE PLACE AND ROLE OF THE DEEP DEPOSIT IN THE STRATEGY OF KGHM "POLSKA MIEDZ" Co.

In the result of socio-political transformations in Poland the state-owned company, Mining & Metallurgical Copper Plant, was transformed on September 9, 1991, into a joint-stock company and named KGHM "Polska Miedz" (source: KGHM website). Its stock was owned entirely by the state treasury.

Until the mid-1990s the strategy of the Company was focused on deep restructuring. At the same time, having in mind the ongoing depletion of ore reserves within the lease areas the Company mined within at the time, the company's Board of Directors undertook a number of initiatives to acquire new deposits. Their main focus at the time was placed on acquiring a deposit outside Poland. In the result, in 1996 the Board made a controversial decision to invest in the "Kimpe" copper-cobalt deposit in Zaire (now Congo DR). This was to be a pilot project, intended to become a springboard for the KGHM's further expansion abroad.

At the same time actions were taken to identify the potential for expanding the company's mineable reserves basing on the mineral resources of the orebody it owned. It was only natural for KGHM to include the idea of developing the deep orebody, bordering to North-West with "Rudna" and "Sieroszowice" mines, into its strategy. With time the importance of the deep deposit grew considerably and the company sought incessantly to acquire the rights to mine it. The first step on that road was to acquire relevant geological information (Banaszak, Tomanik 2002).

With the new Polish "Geological and mining law" enacted (the Act of February 2, 1994) ambiguities occurred as to the ownership of geological reports on "Glogow", "Bytom Odrzanski" and "Retkow" tenements, which until then KGHM regarded as its prospective orebodies. Subsequently in 1996 the Company applied to the Ministry of Natural Resources and Environment to establish ownership status of the reports in question (Banaszak, Tomanik 2002). The Company provided evidence that it directly or indirectly financed the exploration programme, the results of which the geological reports were based on. It was proved that some 60% of the drilling works was financed by Mining & Metallurgical Copper Plant in Lubin.

The main strategic objective of KGHM after it had listed on stock exchange was to ensure that in long-term its value continually grew. The main directions of the Management's efforts were: further restructuring and increasing effectiveness of the Company's strategic assets, consistent implementation of a cost reduction programme and expanding the scope of the Company's business activities to include mining and processing to other metals (Siewierski 1999; Krzeminski 1999). Other strategic goals were to (KGHM 1999, 2000):

- dispose of assets which were not direct stages of the technological process;
- expand into the telecommunications sector;
- invest in orebodies outside Poland (the difficult situation in Africa notwithstanding).

At the same time the Company implemented a strategy of securing itself against the exchange rate risk (KGHM 2000).

Notwithstanding the efforts in other areas, the Company maintained its interest in the deep deposit. In 1997 a report entitled *A variant analysis of the possibility to compile, by KGHM "Polska Miedz" Co., geological reports on the copper-bearing orebodies found below 1,200 metres, considering the geological data and previous reports ownership issues* was compiled. At the same time a decision was made to undertake during 1997–2000 a research project called *Development of the orebody containing copper and secondary minerals occurring in LGOM below 1,200 metres, and copper orebodies in other parts of the country*. In the meantime, thanks to the efforts of the Management, the Minister of the Environment recognized the rights of KGHM "Polska Miedz" Co. to 58.9% of the total value of the geologic information – proportionally to the number of exploratory boreholes owned by the Company (Banaszak, Tomanik 2002).

Fortunately the data from the boreholes owned by KGHM proved to be sufficient for a separate geological report, which scope covered the "Bytom Odrzanski", "Glogow" and "Retkow" areas directly bordering with the orebodies located within the then lease areas, where mining was in progress. Thus it was not necessary for KGHM to buy the remaining 40% of the geological information which, incidentally, the Ministry of the Environment refused to sell.

In December 1998 a geologic report on the copper-silver ore deposit "Glogow Gleboki" compiled by the Polish Geological Institute in Warsaw was approved by the Mineral Resource Committee appointed by the Minister of the Environment. The report was approved by the Minister on March 18, 1999. The document was intended to form the basis for a future mining license application.

Interest in the deep deposit grew. In the years 2000–2001 a research project *Long run development plan for the LGOM copper orebodies (in current and prospective mining areas) including an economic analysis.*

In 2002, together with ABN Amro and J.P. Morgan banks, a document called: *Development strategy of KGHM "Polska Miedz" Co. and the Capital Group for the years 2002–2006 and an outline of the strategy until 2016*, where the following major goals of the Company's strategy were named (KGHM 2003):

- focusing on copper mining and metallurgy,
- increasing value of the existing mining-metallurgical-processing assets,
- expansion in the area of metal mining and processing.

As stated in the document the strategic objective of expansion in the area of metal mining and processing – both within the borders of the then lease areas and outside of them – was to be carried out by:

- mining the orebody which was then outside the lease areas,
- investing in existing mining and mining-metallurgical assets in the world,
- investing in new low-extraction-cost deposits in the world,
- investing in processing activities with synergy potentials.

In view of the above further steps were taken to acquire rights to the deep deposit. Studies, which were essential for the future process of planning of the deep orebody's development, were intensified. At the same time the orebody started to be regarded as very important for Company's future.

Following the period of disputes between KGHM and the Ministry over the issue of ownership of the geological information on the "Bytom Odrzanski", "Glogow" and "Retkow" tenements, a decision was made by the Board in June 2002 to set in motion the mining concession application procedure with respect to the "Glogow Gleboki" orebody. In July 2002 KGHM applied to the Ministry of the Environment for the orebody to be evaluated and royalty payments for the area, for four alternative extraction scenarios, calculated. The evaluation for all four variants was completed in March 2003 and in July of the same year boarders of mining areas applicable to each of the variants were outlined.

Despite KGHM's strong focus on identifying investment possibilities abroad, in South America and Africa among others, and analyses of investment projects it was involved in, 2004 turned out for the deep orebody to be a watershed year. Until then no economic criteria have been defined in the mining law for copper deposits occurring 1,250 metres below the surface, however in January 2004 the Minister of the Environment approved the criteria adopted by the Company for the purpose of documenting the "Glogow Gleboki" orebody. That part of the deep deposit, occurring at depths from 1,100 to 1,400 metres and boardering with "Rudna" and "Polkowice-Sieroszowice" mines, was subsequently named "Glogow Gleboki-Przemyslowy" (Fig. 1, Tab. 1). At the end of the month the Board of Directors of KGHM "Polska Miedz" Co. applied to the Minister of the Environment to provide, for a fee, additional geological data in scope sufficient to cover the area of the first (widest) extraction scenario (Banaszak, Tomanik 2002).

Under the agreement signed in April 2004 between the Minister of the Environment and KGHM, the Company obtained rights to geologic data from 42 additional exploratory boreholes. This allowed work on preparation of geological report for the GGP orebody to be commenced. At the same time other documents required when applying for a mining concession – a development plan for the reserves and an environmental impact report – were being prepared. In April of the same year the location of a new shaft, crucial to the exctraction of the deep orebody, was defined.

In 2004 CBPM "Cuprum" Ltd. completed a development plan and following that a feasibility study for the GGP orebody, allowing a decision to go ahead with development in the area to be made.

In November 2004, under the agreement between KGHM "Polska Miedz" Co. and the State Treasury represented by the Minister of the Environment, a mining lease area for the GGP deposit was established (valid for 50 years from the day of granting a mining concession), and, a day later, the

Minister of the Environment granted KGHM with a concession to mine copper ore from the GGP orebody.

Still in the same year development of the deposit was started from "Rudna" mine.

The year 2005 saw the development continue.

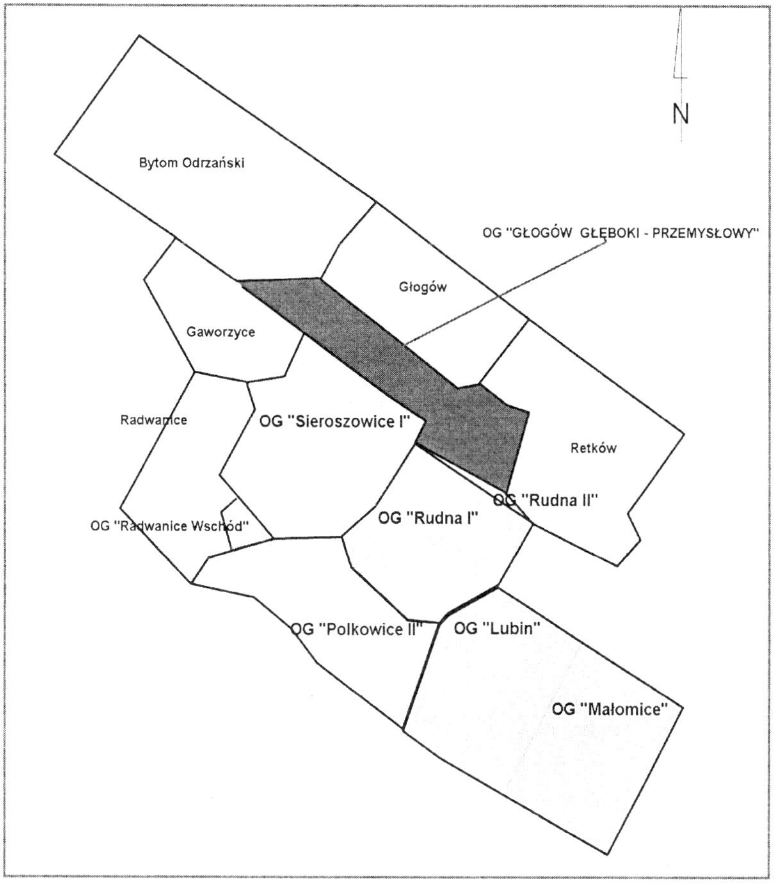

Figure 1. Mining tenements in the Legnica-Glogow area

Table 1. Concession lease areas of the KGHM "Polska Miedz" Co.

Tenement	Acreage [sq. kilometres]
Malomice I	75.6
Lubin I	82.6
Polkowice II	75.7
Radwanowice-Wschod	3.3
Sieroszowice I	97.0
Rudna I	75.6
Rudna II	2.2
GGP	56.0

In his letter dated June 2005 the Director of the State Mining Authority instructed KGHM to present its overall strategy of mining in the GGP area for opinioning by the Commission on Mining Induced Seismicity, Support and Roof Control in Copper Ore Mines. In the result the Commission obligated the Company to:

- provide the Commission with a life-of-mine plan for the extraction of the deposit in question including the development workings planned in the "Glogow Gleboki – Przemyslowy" area;
- assess the widths of safety pillars required to protect access drives to the deposit occurring at depths between 1,100 and 1,300 m.

In July 2006 the Board of Directors of KGHM "Polska Miedz" Co. made a decision to go ahead with the construction of a new shaft, "Glogow Gleboki" (SW4). In October of the same year the "Development strategy of KGHM "Polska Miedz" Co. and the Capital Group for the years 2007–2016" was adopted (KGHM 2006). The document stated that the strategy of KGHM "Polska Miedz" Co. was to focus of increasing the value of the Company by maintaining its position as a world copper and silver producer and expanding into the area of mining and processing of other metals. The document recognized accessing and developing the GGP deposit as one of the main strategic goals of the Company, essencial for expanding its mineral reserve base and maintaining the then level of copper production from domestic orebodies. Developing and mining the GGP reserves and sinking SW4 shaft (1.5 billion PLNs) were regarded – on par with foreign acquisitions and smelter modernization project (1.3 billion PLNs) – as top priority (PLN = Polish zloty; 1 PLN ≈ US$0.3).

In 2007 the construction of a 1,250-metres-deep "R-XI" shaft was completed together with a central cooling system – the first gateway to the GGP reserves. At the same time sinking of SW-4 shaft, an essential access way to the deep deposit was commenced and horizontal drivage continued.

3. DEVELOPMENT SCENARIOS FOR THE GGP AREA

In July 2002 KGHM applied to the Ministry of the Environment for the orebody to be evaluated and royalty payments for the area, for four alternative extraction scenarios, calculated (Fig. 2):

1. variant i – ore reserves of 204.1 million tonnes,
2. variant ii – ore reserves of 29.7 million tonnes,
3. variant iii – ore reserves of 38.5 million tonnes,
4. variant iv – ore reserves of 85.8 million tonnes.

As already mentioned a mining concession was granted for the lease area in variant i.

In 2003 three alternative deposit development scenarios were presented:

1. scenario 1, in which sinking of three shafts: material-and-man hoisting, ore hoisting and ventilation was assumed;
2. scenario 2, in which sinking of three shafts: material-and-man hoisting, ore hoisting and three ventilation was assumed;
3. scenario 3, in which sinking of one ventilation shaft and driving main access ways from "Rudna" and "Polkowice-Sieroszowice" mines was assumed.

In variants 1 and 2 erecting of ore enrichment plants with adequate throughput capacities was assumed. In variant 3 the mined ore was to be processed at "Rudna" mine's plant.

In all scenarios the concentrate was to be dispatched to "Glogow" smelter and the flotation waste was to be deposited in "Zelazny Most" ("The Iron Bridge") waste storage dam.

The basic differences between the variants resulted from differences in the assumed capital expenditures, which determined production levels in each case, mainly due to different numbers of ventilation shafts.

In the case of scenario 1, the total capital expenditure was assumed to amount to ca. 3.1 billion PLNs, of which shafts "GG-1"/"GG-2" were to cost ca. 1,039 million PLNs, underground infrastructure – 903.9 million PLNs, a copper ore processing plant – 548.6 PLNs million, sinking and construction of "GG-3" shaft –407.6 milllion PLNs.

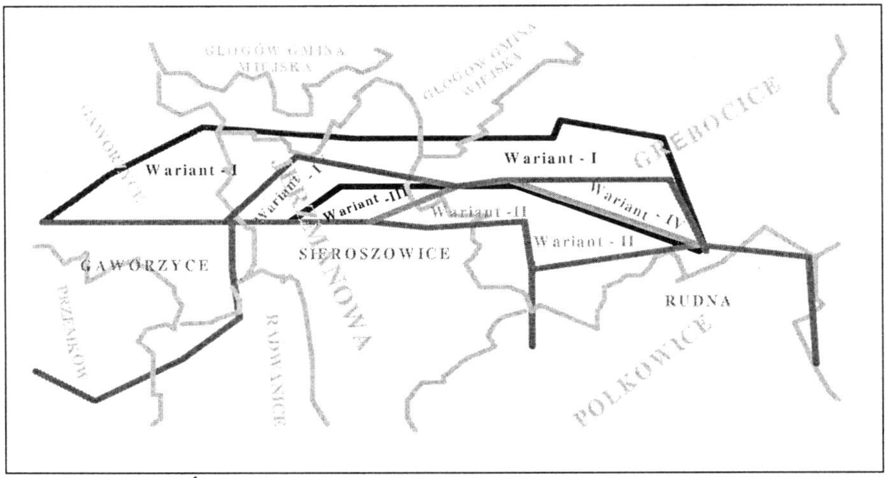

Figure 2. Alternatives of mining tenement areas

It was estimated that in the case of adopting scenario 1 the project would take 16 years to complete, starting in 2004.

The total volume of air delivered to the new mine through "GG-1" and "GG-2" downcast shafts should not exceed 70 thous. cubic metres. Average daily production was planned at 18 thous. Tonnes/d (5.2 million tonnes per annum).

In the case of scenario 2, the total capital expenditure was assumed to amount to ca. 4.4 billion PLNs, of which shafts "GG-1"/"GG-2"/"GG-4" were to cost ca. 1,357.4 PLNs million, underground infrastructure – 1,063.4 million PLNs, sinking of "GG-3" and "GG-5" shafts – 407.6 million PLNs each, a copper ore processing plant blocks 1–2 and 3 – 548.6 and 396.6 milion PLNs respectively.

It was estimated that in the case of adopting scenario 2 the project would take 20 years to complete (2004–2023).

The target production rate in scenario 2 was calculated as 32 thous. tonnes/d (9.3 million tonnes per annum).

In the case of scenario 3, the total capital expenditure was assumed to amount to ca. 1.3 billion PLNs, of which shaft "GG-1" was to cost ca. 380 million PLNs and underground infrastructure – ca. 848 million PLNs.

It was estimated that in the case of adopting scenario 3 the project would take 16 years to complete. Access drives were to be driven starting in 2004.

The target production rate for scenario 3 was calculated as 15 thous. tonnes of ore per day (4.35 million tonnes per annum). It was expected to extend the life of the mine until the end of 2040. According to the plan the first ore from production faces was to be mined in 2014 and the target production rate was to be achieved in 2019.

Due to the fact that scenarios 1 and 2 turned out not to be economically viable, it was scenario 3 which was accepted to form the basis of the ruling mine design, which was described in *A feasibility study of extracting copper ore in the "Glogow Gleboki – Przemyslowy" area – technical and economic assumptions* (KGHM website). In the study it was assumed that in long-term all KGHM mines would be combined into one organism. It was further assumed that the optimal development of the GGP orebody would be assured by creating ventilation and transport systems, which would be common for the whole deposit and that rational and maximal utilization of its mineral reserves, would require full co-ordination as far as organization and technology are concerned.

Studies and research showed that a separate mining in the GGP tenement would be uneconomic so that a model of so called "united mine", was adopted. It is assumed, that this super-mine, being a union of existing mines, will produce ca. 29–30 thous. tonnes/d. According to this concept, the GGP orebody is being developed from "Sieroszowice" and "Rudna II" tenements. By 2020, KGHM mines should drive over 700 thous. metres of tunnels (60–70 kilometres yearly). The Board presumes that in this way all the existing shafts and tunnels will be exploited with maximal use. Due to high temperature, exceeding 45°C, a suitable connection between shafts, ensuring sufficient fresh air supply, is planned.

CONCLUSION

Until the decision to develop the GGP deposit was made in July 2006, KGHM "Polska Miedz" Co. held concessions to mine reserves in the following tenements (Fig. 1):

1. "Lubin-Malomice",
2. "Polkowice",
3. "Sieroszowice",
4. "Rudna" ("Rudna I" and "Rudna II" mining areas),
5. "Radwanice-Wschod".

The concessions entitled KGHM to mine copper ore: in "Lubin-Malomice", "Polkowice", "Sieroszowice" and "Rudna I" mining tenements – until 2013, "Radwanice-Wschod" mining area – until 2015, and in "Rudna II" – until 2016.

It was obvious that in order to maintain the then production level the Company would have to avail itself of other mineral reserves, in the country or abroad. In March 2000 the Mineral Resource Department of KGHM "Polska Miedz" Co. conducted an analysis in order to assess the probable time when the ore reserves within the existing lease areas would be exhausted, considering the economic viability of their extraction.

It was only natural that KGHM showed interest in the deep orebody, which was an extension of the deposits mined currently at "Rudna" and "Sieroszowice" mines and bordered with them to North-West. With time, as the Company's ore reserves were gradually being exhausted, mining progresssed deeper and technological advancements were made, the importance of the deep deposit in KGHM's strategy grew considerably.

Exploration results showed that the deep deposit contained ore whose quality, in respect of copper and silver grades, was comparable to the rich ores mined in some areas at "Rudna" mine. Unfortunately the overall level of geological knowledge of the deep deposit is not satisfactory. Moreover the geothermic gradient in the area is unfavourable and high rock temperatures are expected at mining faces located in the "Glogow Gleboki" deposit area. On average in the Fore-Sudetic Monocline for every 100 metres increase in depth temperature increases by 3°C. Currently at 1,000 metres below the surface virgin rock temperature reaches 33°C. Another 300–400 metres of depth translates to additional 10–12°C of rock temperature. Meanwhile the mining law permits work of people in ambient temperatures not higher than 28°C. This implies that mining in the area would mean high costs of cooling. Large depths also mean increased seismic hazard, which also translates to higher costs. It must only be hoped that all these problems will be overcome. Considering all the mining and geological conditions – including mining width – they assume to use the verified mining methods. With progress of extraction mining methods will be modified.

What must also be kept in mind is that below the GGP, "Glogow", "Bytom Odrzanski" and "Retkow" tenements are located where copper mineralization was discovered. These orebodies extend down to 1,500 m below the surface. Because no economic criteria for copper ore deposits occurring at depths below 1,250 metres are currently available mineral reserves in these areas have not as yet been documented. It can be presumed that KGHM will also have interest in these tenements.

REFERENCES

Banaszak A., Tomanik R. 2002: Formalnoprawne aspekty zagospodarowania złoża "Głogów Głęboki" ("Formal & Legal Facets of Development of "Glogow Gleboki – Przemyslowy" Copper Deposit"). Mat. konf. "Wykorzystanie zasobów złóż kopalin użytecznych", wyd. IGSMiE PAN.

KGHM wesbsite: Historia („History"). Strony internetowe KGHM „Polska Miedź" S.A. www.kghm.pl

KGHM 1999: Korekta programu rozwoju KGHM "Polska Miedź" S.A. na lata 1999–2003 ("Correction of KGHM "Polish Copper" Development Programme"). Lubin, lipiec 1999.

KGHM 2000: Sprawozdanie Zarządu z działalności Spółki w 1999 roku (KGHM's Board Report for the Year 1999). KGHM "Polska Miedź" S.A., Lubin, marzec 2000, s. 16.

KGHM 2003: Strategia rozwoju KGHM "Polska Miedź" S.A. i Grupy Kapitałowej na lata 2002–2006 oraz założenia do strategii do roku 2016 ("Development Strategy of KGHM "Polska Miedz" Co. and the Capital Group for the Years 2002–2006 and an Outline of the Strategy Until 2016). Mat. KGHM "Polska Miedź" S.A., 2003.

KGHM 2006: Strategia KGHM "Polska Miedź" S.A. i Grupy Kapitałowej na lata 2007–2016 ("Development Strategy of KGHM "Polska Miedz" Co. and the Capital Group for the Years 2007–2016). Mat. KGHM "Polska Miedź" S.A., 2006.

Krzemiński M. 1999: List do akcjonariuszy ("A Letter to Shareholders"). Dokum. KGHM "Polska Miedź" S.A.

Rydzewski A. 1996: Krótka historia poszukiwań i odkrycia złoża miedzi na monoklinie przedsudeckiej ("A Brief History of Exploration and Copper Deposit Discovering in the Fore-Sudetic Monocline"). Wiadomości PIG, 1996.11, nr 63.

Siewierski S. 1999: List do akcjonariuszy ("A Letter to Shareholders"). Dokum. KGHM "Polska Miedź" S.A.

International Mining Forum 2009, Sobczyk, Kicki & Saługa (eds) © 2009 Taylor & Francis Group, London, UK, ISBN 978-0-415-80428-8

The Prognosis of the Climatic Hazard for the Growing Depth of Underground Mining

Krzysztof Słota
Silesian University of Technology, Gliwice, Poland

SUMMARY: Together with the growth of depth of underground mining and the progressing concentration of production grows the climatic hazard. Early prognosis of this threat permits to choose suitable techniques and air-conditioning methods already in the stage of designing the exploitation of the deposit. In this work it is shown how the primal temperature of rock changes and the climatic hazard rises with the growth of depth of underground working. It is also explained how the polish regulations define "climatic hazard", the methods of improvement of the climatic conditions are introduced as well as is the prognosis of demand on cooling power.

KEYWORDS: Climatic hazard, ventilation, air-conditioning

1. INTRODUCTION

In Polish coal and copper mining industry grows the depth of underground working as well as the concentration of production, which causes, inter alia, the growth of climatic hazard. The depth of exploitation in mineral coal mines increases yearly in about 8 m [13]. Taking the geothermal gradient into account (33 m/°C for GZW and LZW, 28 m/°C for a copper basin, and in case of anomaly in JSW 24–27 m/°C) it is clear, that we are dealing with a situation, in which the new excavations (every 3–4 years) are conducted amidst rocks with their primal temperature elevated by approximately 1°C. Also the machines and devices, with their power still growing, which are used in the process of excavating, have the influence on the scale of the climatic hazard.

The statistics show, that over 1/3 of the yield of mineral coal and copper ore is extracted in excavations, in which the DBT is higher than 28°C. Every 24 hours circa 10 500 workers dwell in the harsh conditions of the microclimate. It is bound for the number of excavations with dry temperature exceeding 28°C and the number of their workers to increase in the upcoming years.

In the year 2007 quarrying processes were conducted in 30 mines with 127 longwalls, in 27 of those mines excavations were carried out with 75 longwalls below the level of accessibleness. The dominating system of aeration was the so called "U" system, which was used in 98 longwalls. The "Z" system aerated 8 of them, while the "Y" did so with 19. In the longwalls, which were conducting quarrying processes below the level of accessibleness: 5 of the longwalls were managed on a depth not exceeding 500 m, 39 longwalls on a depth between 500 and 800 m, and 31 longwalls below 800 m. 45 longwalls were managed with the difference between the depth of accessibleness and the depth of the conducted work greater than 50 m, amongst them 25 longwalls were managed with the difference greater than 100 m. The exceedance of 28°C in dry temperature was stated in 12 longwalls, in 37 longwalls the dry temperature amounted between 25 and 28°C.

Table 1. The condition of the crew employed in a shortened amount of working time [13]

The condition of the crew employed in a shortened amount of working time			
Year	The number of coal mines	The number of copper mines	Number employed
1997	26	3	5160
1998	22	3	4929
1999	15	3	4179
2000	20	3	4815
2001	20	3	5676
2002	21	3	6364
2003	21	3	6019
2004	27	3	6709
2005	22	3	6811
2006	21	3	7328
2007	18	3	10511

At present the primal temperature of the rock reaches, in some mines, 45°C and it is to be expected, that in the future the excavations will be conducted in a rock with its primal temperature exceeding 50°C. At that time the climatic hazard will be one of the primary factors endangering the miners and the possibility to continue their work.

An early prognosis on the climatic hazard, which should be carried out at the stage of designing the process of quarrying the ore deposit. This will permit to choose relevant ventilation and air-conditioning techniques and methods, which will enable the miners to continue the underground working.

2. CLIMATIC HAZARD

The proper recognition of the processes of heat and moisture exchange in the underground mining has an essential meaning in defining the scale of the climatic hazard. The information about the hazard is needed to adequately design the method of creating access to the ore body, managing the underground mining, choosing the technological process and establishing the safe conditions of work (inter alia the work time, the energy expenditure and the heat stress).

The basic parameters (e.g. temperature, moisture, velocity) of air and surroundings are used alongside with the so called microclimatic indexes (e.g. WBGT, ATE, t_{zk}, δ) to define the state of climatic hazard in the mines.

The value of the primal temperature of rocks on the deepest level of excavation is the measure of climatic hazard in the whole mine (on Figures 1 and 2 the isolines of the primal temperature of rock for JSW S.A. were shown; those were made for the horizons amounting between –650 m and –850 m [8]). All mines were divided into four groups, depending on the value of the earlier mentioned temperature:

- The first group features the mines, in which the primal temperature of rocks on the deepest exploitional level is greater than 40°C. Those are the mines with a very high level of climatic hazard.
- The second group features the mines, in which the primal temperature of rocks on the deepest exploitional level is greater than 35°C but does not exceed 40°C. Those are the mines with a high level of climatic hazard.

- The third group features the mines, in which the primal temperature of rocks on the deepest exploitional level is greater than 30°C but does not exceed 35°C. Those are the mines with a low level of climatic hazard.
- The fourth group features the mines not exposed to the climatic hazard. Those are the mines, in which the primal temperature of rocks on the deepest exploitional level is lower than 30°C.

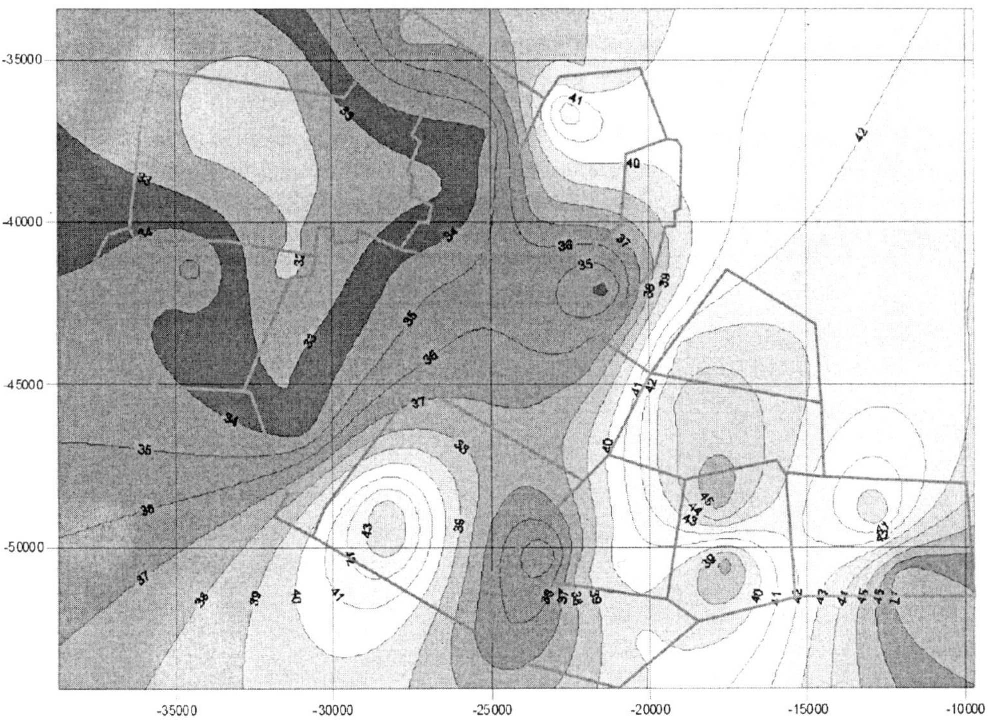

Figure 1. The map of the isolines of the primal temperature of the rock for the horizon amounting –650 m (for JSW S.A. [8])

The climatic index "K" is the measure of the climatic hazard in reference to the exploitional level. It is defined by the equation written below [1]:

$$K = \frac{t_{pg} - t_d}{t_d - t_p} \tag{1}$$

where: t_{pg} – the primal temperature of rocks on a specified level, °C; t_d – admissible temperature of air at working place (without the applying of a shortened time of work), °C; t_p – the temperature of air in the cowl section of the level, from which fresh air is administered to the exploitional and preparatory excavations, °C.

By interpreting the value of the above mentioned index, we can define the scale of the climatic hazard:
- $K < 0$ – no climatic hazard;
- $0 \le K < 0,8$ – there is a small climatic hazard, which can be prevented by obeying the main rules of rational ventilation;

- $0{,}8 \leq K < 1{,}5$ – there is a high possibility of the existence of a climatic hazard, which should be dealt with by using the ventilation methods of climatic hazard prevention, and in case these would turn out insufficient, by using the cooling devices;
- $K \geq 1{,}5$ – there is a very high possibility of the existence of a climatic hazard, which should be dealt with by using the cooling devices.

This index is used only to pre-evaluate the possibility of existence of a climatic hazard. It is like that, because there have been cases of excavations located on the levels with a very high climatic hazard index ($K > 1{,}5$), where the temperature was much lower than 28°C. This has been reported to work both ways, meaning that on the levels with no climatic hazard whatsoever ($0 \leq K < 0{,}8$) excavations with the air temperature exceeding 28°C existed.

The most precise evaluation of safety level, in reference to the excavations being designed, is achieved by basing on the weather forecasts.

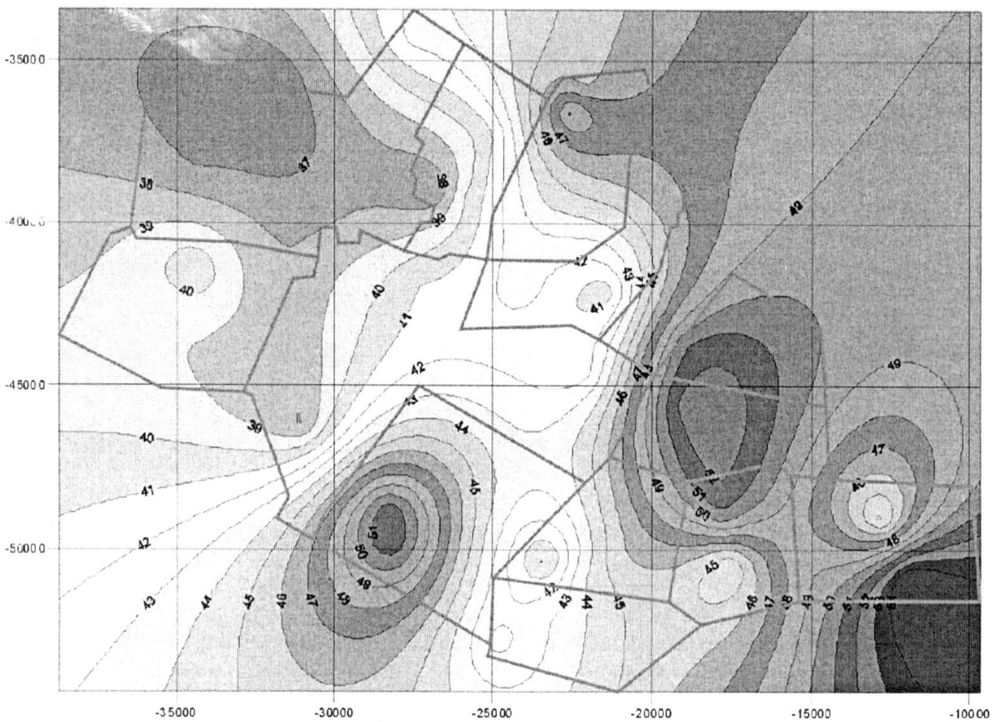

Figure 2. The map of the isolines of the primal temperature of the rock for the horizon amounting –850 m (for JSW S.A. [8])

According to the recent mining regulations [14], the dry temperature value and the intensity of cooling measured with a moist katathermometer are the measure of the climatic hazard, in reference to the existing working places.

In the regulations, which are currently mandatory for the Polish mining enterprises, it is set that: at the working place the dry temperature should not be higher than 28°C and the cooling intensity should not be lower than 11 kata; in case the air dry temperature is greater than 28°C, but does not exceed 33°C or the cooling intensity is lower than 11 kata, technical solutions should be applied in order to lower the dry temperature value or the work time should be shortened. In case, when the dry temperature exceeds the value of 33°C, only search and rescue operations carried out.

With the prima temperature of the rock higher than 30°C, a prognosis on the climatic conditions should be worked out alongside with the methods to maintain the accurate ones.

In the mining enterprises, which use self-propelled machines, it is possible to define the climatic conditions of work, by determining the equivalent temperature of climate in the way administered by the Polish Standard.

The mining manager calls a climatic hazard team up, which is supposed to analyse and evaluate the state of climatic hazard in the mine on the basis of: defining the primal temperature of the rock, the natural conditions of the overlie, mining conditions in the existing batches of deposit and those being designed, the actual ventilation measurements with the allowing of the microclimate (temperature, intensity of airing and cooling and the relative humidity) as well as the power of the set up mechanical devices.

The team also deals with the current evaluation of the implementation and efficiency of the new introduced methods and ways to control the temperature hazard.

3. THE WAYS OF IMPROVING THE CLIMATIC CONDITIONS

There are five ways of improving the climatic conditions at working places and in people's gathering places:

- Lowering the temperature (air-conditioning).
- Decreasing the dampness of air (air dehumidification).
- Increasing the velocity of the flow of air.
- Lowering the mean of the radiation temperature of the surroundings.
- Decreasing the mean of energy displacement of the workers during the shift by creating a rest break every hour of the working day.

The lowering of the air temperature causes the amplification of the heat emanation through convection.

The decreasing of the humidity of air increases the evaporation of sweat.

The growth of the velocity of air makes it possible to increase the emanation of heat through convection as well as the evaporation of sweat. The lowering of the mean of the radiation temperature of the surroundings uses the radiation of warmth of the human body to transport the warmth of metabolism. Decreasing the energy displacement of the human being causes the reduction of the produced metabolism energy, which is also the reason of the improvement of the climatic conditions and the liquidation of a dangerous for health and life accumulation of warmth.

Different air-cooling (air-conditioning) methods are used in polish and world mining. The following types of air-conditioning were distinguished (the division with regard to the place of mine air-conditioning):

In world mining and Polish the different methods of refreshing air be applied (air-conditioning). It the following arrangements of air-conditioning were distinguished was (the division with regard on place of air-conditioning of mine air):

- central air-conditioning;
- with the central system of cooling the air on surface;
- with central underground system of cooling the air of shaft bottom;
- group air-conditioning – with an underground system of cooling grouped currents of air;
- local air-conditioning – the cooling of the air with the use of the local air-conditioning appliances placed in the provincial air currents.

The mentioned air-conditioning systems have good, as well as bad, sides and the usage of a specific system depends, above all, on the cooling power requirement, the technical and organisational capability, the ventilation conditions of the mine and also on the reimbursement There is also a possibility to mix different air-conditioning systems in order to improve the climatic conditions. The air-conditioning devices for the cooling of the air in the heading are used in Polish mining from many years. At present those have the value of the cooling power varying from 30 to 600 kW, the products of WAT, Wende-Malter, Termospec, IMKiUS (which stands for the Engineering of the Climatic Machines and Special Devices) as well as Mostostal Wrocław.

The devices with smaller power are used in the longwall excavations, those are the longwall air coolers, which are mainly powered by the water-cooling aggregates (cooling power between 30 and 60 kW) situated usually at the top road. Machines with the cooling power between 250 and 350 kW functioning directly or indirectly are the air-conditioning appliances cooling the air in provincial currents. The efficiency of the devices depends on their technical state as well as on the parameters and the quantity of the cooled air. The fans, which force the air to flow through the coolers, have an essential significance in the cooling of the air with the use of air-conditioning devices. Their work can cause the temperature of the air flowing through the cooler to increase even for about 5°C, which leads to the loss of heat that can reach up to 40 kW.

4. THE PROGNOSIS ON THE CLIMATIC CONDITIONS AND THE COOLING POWER REQUIREMENT

Numerous works and the studies of methods of the prognosis on climatic conditions [1], [2], [3], [4], [6], [7], [9], [10], [11], [12] describe the processes of warmth and humidity exchange in the excavation. They are based on the assumption that the rock mass is homogeneous and isotropic. The individual methods differ between themselves in the way of describing the supervening processes of warmth and mass exchange between the surroundings and the flowing air in the excavation (also in the longwall heading). The mathematical study of the mentioned case boil down in their final form to the administration of the correlations defining the final parameters o fair – mainly the temperature and humidity. The functions o these parameters depend on the initial parameters of the air at the entry to the heading as well as on the individual sources of heat and humidity of that place. The recognition and a detailed description of all of the possible to define heat and humidity sources, existing in the excavation, are crucial for the accurate prediction of the climatic conditions in the longwall excavation.

The admission of heat from the rock, from the oxidising carbon, from the machines and devices, from the transported excavated material, the admission of the latent heat along with the water vapour from the rock, the heat off take through the cooler, the capture of air heat by the cold water pipeline, the evaporating of water out of the open areas of the excavation, the admission of water vapour from the rock, the phase transitions of the water contained in the air, the condensation of water vapour in the coolers, the sources of the water vapour associated with the local positive heat sources – these factors are counted among the crucial heat and humidity sources.

4.1. *The method of predicting the parameters (temperature, specific humidity) of air flowing through the site of mining by using the longwall air coolers*

In the model of heat and humidity exchange it is assumed that the climatic conditions are accurately described by the temperature of air (t) measured with a dry bulb and by the specific humidity (x). The above mentioned parameters (functions of these parameters) depend on the spatial co-ordinates as well as time. Time (τ) and one current co-ordinate measured alongside the axis (s) of the excavation remain independent variables, while assuming the univariativeness of the flow. The model takes the cooling of air with longwall air cooler into account (with well-known: position of the coolers in the excavation, their power, the kinds of ventilators used and the value of cooled air stream capacity). Below a simplified algorithm for predicting the parameters of air flowing through the site of mining with the use of longwall air coolers, is introduced. In the algorithm it is assumed, that the places of location of the longwall air coolers are known. It is also accepted, that the cooling powers of the longwall air coolers, the streams of cooled air capacity and the kind of ventilators extorting the flow of air through the coolers, are known. The model introduced below can also be used for predicting the climatic conditions in the longwall excavation without the use of air cooling. The algorithm was applied in a prognostic programme of the Z.K.S. Mopfit 2000 [4], which is used in polish coal mining for predicting temperature and humidity in the longwall heading.

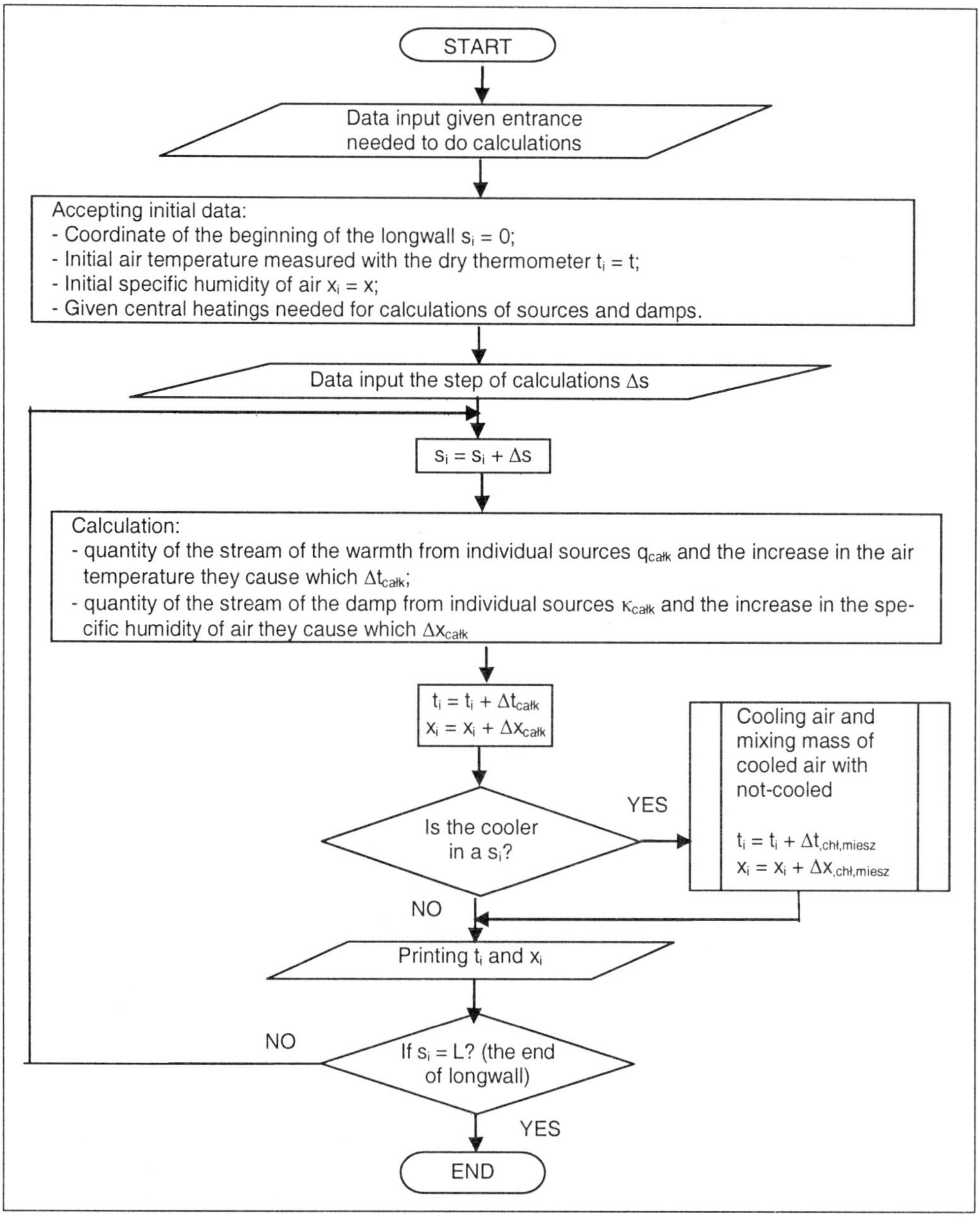

Figure 3. The simplified computational algorithm

Three methods o fair-conditioning of the longwall excavations in polish mining are introduced below (Fig. 4).

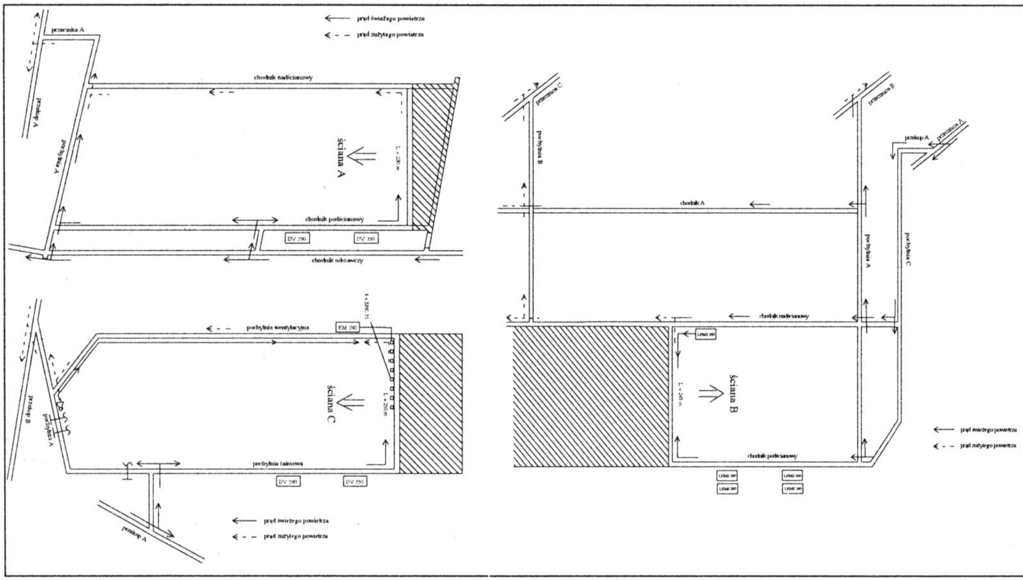

Figure 4. Regional ventilation of longwalls A, B and C

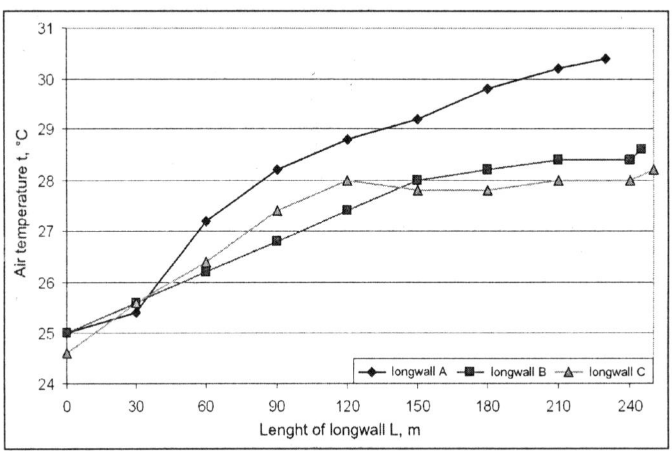

Figure 5. The comparison of the air-conditioning systems –
the changes of the dry temperature value alongside the longwall

In everyone of the ventilation regions of those three longwalls a local air-conditioning, based on the cooling of air in the inlet heading of the longwall-hole, was used. It featured the usage of direct action machines, such as: LKM-2 290, DV-290 and DV-350. In the first longwall working (longwall A), the air was cooled only in the bottom road – that is the traditional and most commonly used air-conditioning system of longwall excavations – the so called deep cooling. In the second longwall heading (longwall B) the air was cooled in the bottom road and was transported to the upper part of the longwall by a pipeline connected with an air-conditioning appliance situated in the top road. In the third one (longwall C) the air was cooled in an inclined transport plane and also by the eight air coolers powered by the cold water from the aggregate situated in the inclined ventilation plane – this is a typical example of a multistage air-conditioning. Figure 5 shows a change in the dry temperature value alongside the longwall excavation for the considered workings.

4.2. *Defining the demand for cooling power*

Using the already existing results of measurements of climatic conditions (mainly the temperature of air) in the exploited longwall excavations and the driven roads as well as the results of the air temperature predictions, it is possible to define the potential demand for cooling power for the ventilation regions with quarter periods division, falling on the individual mining levels. The values of cooling power mentioned in the article define the quantity of heat, which needs to be taken from the longwall area in the way of the admission of fresh air, in order to maintain the accurate climatic conditions defined by the regulations.

The paper introduced the conception of local air-conditioning for the 750 m level and group air-conditioning for 950 m level (figure 6) formulated by the author and approved, on the basis of research and analyses carried out in coal mine "Sośnica-Makoszowy" Movement "Sośnica" [5]. The conception itself regarding to:

- the mining and geological conditions of the present and future exploited longwall excavations and driven roads;
- the ventilation capabilities of mines – the way and the quantity of distributed air;
- the planned exploitation, which relates to managing three or four longwalls simultaneously;
- the assumed daily excavation reaching the level of 10000 tons.

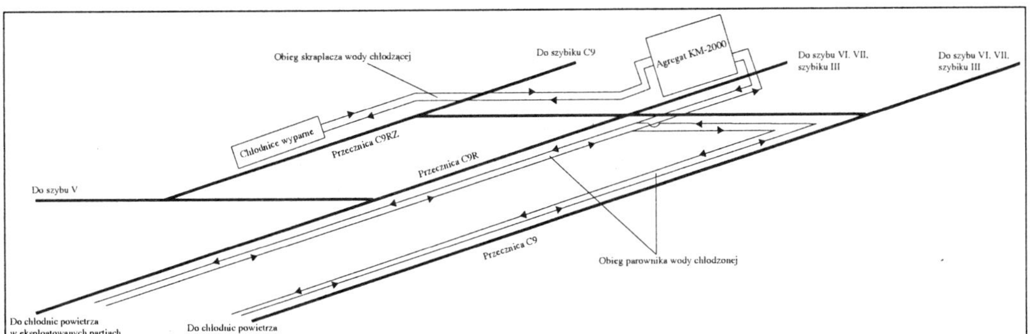

Figure 6. The diagram of group air-conditioning connections

The potential demand on cooling power for exploitional excavations and preparatory workings on the 750 m level is shown in Figure 7.

Figure 8 holds the information regarding the 950 m level. The complete potential demand on cooling power for the longwalls and the preparatory workings of the "Sośnica's" Movement is shown in Figure 9.

The research and analyses conducted, point out that in order to secure the proper working conditions for the work carried out on 750 m level, the required cooling power is one of the value between 0 and 1500 kW.

Since the year 2006 the average demand on cooling power will equal about 800 kW, therefore the use of local air-conditioning for air-conditioning purposes should suffice (for example: 4 machines like DV-290/TS-300 or four DV-150 machines and two DV-290, with the condition that the smaller cooling devices would be used for preparatory workings). For air-conditioning of the headings on 950 m level the usage of group air-conditioning, consisting of e.g. a KM-2000 aggregate (with it's cooling power reaching 1750 kW) or a KM-3000 aggregate (with the cooling power of 2600 kW) and an amount of air coolers with equal cooling power (e.g. four coolers with the power of 290 kW and four with the power of 150 kW for the KM-2000), is advised.

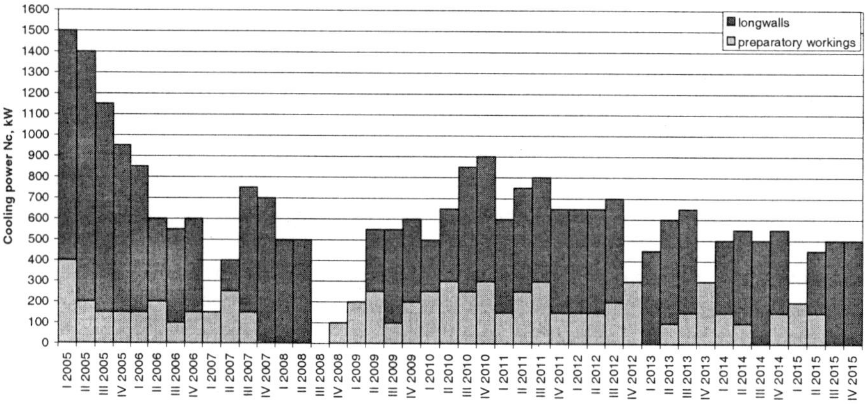

Figure 7. The potential demand on cooling power for longwalls and preparatory workings on the 750 m level

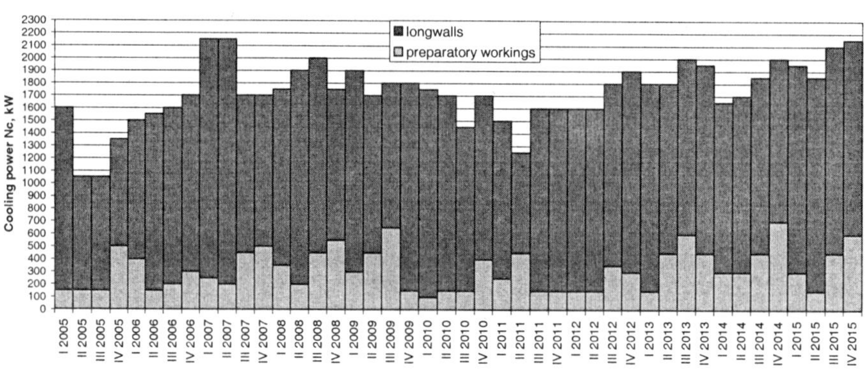

Figure 8. The potential demand on cooling power for longwalls and preparatory workings on the 950 m level

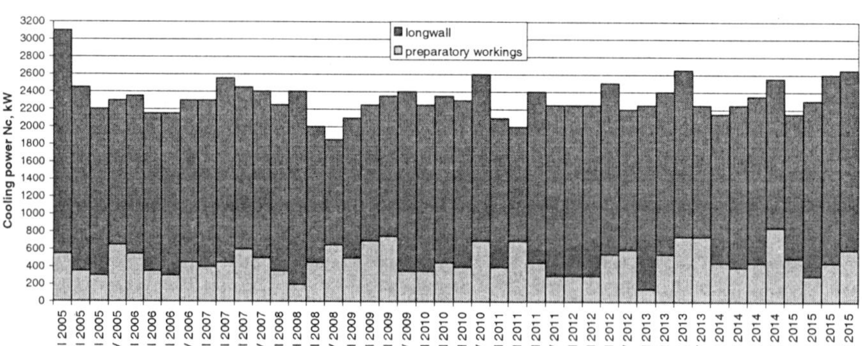

Figure 9. The total potential demand on cooling power for longwalls and preparatory workings

REFERENCES

[1] Cygankiewicz J., Knechtel J., Wacławik J.: Czynniki kształtujące warunki klimatyczne w wyrobiskach eksploatacyjnych. Archiwum Górnictwa, Vol. 39. Warszawa – Kraków 1994.

[2] Filek K., Nowak B.: Temperatura i wilgotność chłodzonego powietrza w podziemnych wyrobiskach górniczych przewietrzanych wentylacją opływową. Rozprawy Monografie 48, Wyd. AGH, Kraków.

[3] Frycz A., Gumiński A., Kolarczyk M.: Procesy wymiany ciepła i wilgoci w ścianach o wysokiej koncentracji wydobycia. XXIV Dni Techniki ROW '98, Seminarium nt. "Zwalczanie zagrożenia klimatycznego w kopalniach – teoria i praktyka", Wodzisław Śląski.

[4] Słota K.: Zmiana temperatury i wilgotności właściwej powietrza przepływającego przez ścianę prowadzoną w trudnych warunkach klimatycznych. III Szkoła Aerologii Górniczej, str. 473–484. Wyd. EMAG, Kraków 2004.

[5] Słota K., Słota Z., Świda I.: Zagrożenie temperaturowe oraz koncepcja klimatyzacji grupowej w KWK "Sośnica-Makoszowy" Ruch "Sośnica". Materiały konferencyjne XXXI Dni Techniki ROP 2005, Politechnika Śląska, Instytut Eksploatacji Złóż, str. 47–52, Rybnik 2005.

[6] Strumiński A., Strumińska B.: Zastosowanie gradientów temperatury powietrza do prognozowania mikroklimatu w podziemnych wyrobiskach górniczych. IX Sesja IBMT, str. 95–108, Gliwice 2000.

[7] Szczerbań A.N., Kremniew O.A., Żurawienko W.J.: Rukowodstwo po riegulirowaniju tiepłowowo rieżima szacht. Niedra, Moskwa 1997.

[8] Szlązak N., Tor A., Jakubów A.: Metody zwalczania zagrożenia temperaturowego w kopalniach Jastrzębskiej Spółki Węglowej. Materiały konferencyjne XXXI Dni Techniki ROP 2005, Politechnika Śląska, Instytut Eksploatacji Złóż, str. 63–82, Rybnik 2005.

[9] Voss J.: Grubenklima. Glückauf-Betriebsbücher, Band 27. Essen 1981.

[10] Wacławik J.: Prognoza kopalnianych warunków klimatycznych z uwzględnieniem wpływu wilgotności. Zeszyty Naukowe AGH, seria Górnictwo, zeszyt 33, Kraków 1971.

[11] Wacławik J.: Wpływ temperatury skał na temperaturę powietrza. Rudy i metale nieżelazne, nr 10/1965.

[12] Woropajew A.F.: Tieorija tiepłoobmienia rudnicznowo wozducha i gornych parod w głubokich szachtach. Niedra, Moskwa 1966.

[13] Raport WUG: Zagrożenie wentylacyjno-klimatyczne. Wyd. WUG, Katowice 2008.

[14] Rozporządzenie Ministra Gospodarki z dnia 28 czerwca 2002 r. w sprawie bezpieczeństwa i higieny pracy, prowadzenia ruchu oraz specjalistycznego zabezpieczenia przeciwpożarowego w podziemnych zakładach górniczych.

International Mining Forum 2009, Sobczyk, Kicki & Saługa (eds) © 2009 Taylor & Francis Group, London, UK, ISBN 978-0-415-80428-8

An Assessment of Prospective Reserves for Underground Coal Gasification in Poland

Barbara Białecka
Central Mining Institute, Katowice, Poland

ABSTRACT: One of prospective methods of coal utilization, especially in case of coal resources which are not mineable by means of conventional methods, is underground coal gasification (UCG). This technology allows to recover coal energy "in situ" and thus avoid the health and safety risks related to people which are inseparable from traditional coal extraction techniques.

Prerequisites for the application of hard coal gasification technology include:

- selection of criteria that determine suitability of coal beds for gasification;
- assessment, based on the assumed criteria, of prospective beds for underground gasification;
- specification of hard coal and lignite resources which can constitute the raw material basis for the gasification process.

An analysis and classification of Polish coal resources according to these criteria constitutes the subject of this paper.

INTRODUCTION

In recent years, the global economy has shown a tendency to return to fossil fuels as a primary source of energy. The intrinsically economic coal resources in Europe, which constitute about 25% of world resources of this mineral (Burton, Friedmann and Upadhye 2007), amount to approximately 231 billion metric tons. The largest coal producers in Europe include: the Czech Republic, Germany, Great Britain, Poland, Russia and Ukraine. A substantial part of the solid fossil fuel resources of the European Union, including the ones that are unrecoverable, are located in Poland. This implies that these Polish resources may be important for strengthening the continent's energy security (Darski, Kicki and Sobczyk 2001; PARGWK S.A. 2005).

Being a relatively cheap fuel, coal is a viable option to ensure energy security. Its future, however, largely depends on the application of appropriate technologies that reduce the negative impact of coal combustion on the environment.

Underground coal gasification (UCG) is an appropriate technology to economically access the energy resources in deep or unmineable coal seams and potentially to extract these reserves through the production of gas for power generation, production of liquid fuels or chemicals. Because underground coal gasification (UCG) reduces the emissions of such pollutants as mercury, sulphur and nitrous oxides, it should be considered as one of the technologies enabling the future utilization of coal for energy purposes. Poland is a potentially good area for underground coal gasification.

Coal gasification remains in line with the development strategies of the Polish energy sector, including the policies concerning activities in the field of mining and heavy industry restructuring. Tasks involving the gasification of solid fossil fuels cannot be separated from the development of carbon dioxide sequestration technologies and environmental protection activities in the area of minerals extraction and processing.

Numerous Polish coal mines have substantial hard coal resources; their extraction has stopped due to technical and economic reasons and mining safety issues. These resources include coal deposits located in tectonic disturbance zones, safety berms, prospective and undeveloped deposits as well as thin seams and coal beds of high ash and sulphur content. Only some of these resources can be used with underground gasification methods because of their technical, geological, and mining conditions, such as considerable water inflows, complex macro- and micro-tectonics or high susceptibility to roof falls, which can hinder the continuity and control of the gasification process. Therefore, there is an urgent need to balance the potential resources suitable for UCG according to a set of precisely defined criteria.

Prerequisites for the application of hard coal gasification technology include:

- selection of criteria that determine suitability of coal beds for gasification;
- assessment, based on the assumed criteria, of prospective beds for underground gasification;
- specification of hard coal and lignite resources which can constitute the raw material basis for the gasification process.

In principle, natural factors play a decisive role among the criteria of seam selection for UCG (Blinderman 2005a; Blinderman 2005a; Jaskulski and Rabsztyn 1982). These factors include the coal type and physicochemical properties, depth of occurrence, thickness and angle of inclination of the coal bed, type and tightness of the rock mass, hydro-geological conditions, deposit tectonics, quantity of resources, presence of methane in the deposit and the conditions of the infrastructure on the surface. An analysis and classification of Polish hard coal and lignite resources according to these criteria constitutes the subject of this article.

CRITERIA OF COAL BED CLASSIFICATION FOR UNDERGROUND GASIFICATION

Coal gasification in the bed is a complicated process which, apart from technological factors, is influenced by several natural factors (Tab. 1).

Table 1. Characteristics of classification criteria of coal for underground gasification

Criterion	Characteristics/remarks
coal type	any
physicochemical properties of coal	recommended: high content of volatile matter, low agglomerating capacity or its lack, ash content < 50% by weight
occurrence depth	profitability criterion
bed thickness	more than 1 m
angle of inclination of coal bed	any
type and tightness of rock mass	recommended: firmness and tightness of rock mass, thickness and lithology of rock mass – overburden in slightly permeable layers (clays, silts, shale clays)
Hydro-geological conditions	recommended: lack of fissures, faults, aquiferous layers, water reservoirs causing water inflow
deposit tectonics	recommended homogeneity of deposit (lack of fissure, faults)
quantity of resources	profitability criterion
methane presence in the bed	causes gas hazard
conditions of infrastructure	recommended lack of building development

1) *For underground coal gasification, coals of different type, including brown coals through hard coals and anthracite can be utilized.* Because of the simplicity of creating underground gas generators, most operations in the world (Rauk 1979; Blinderman 2005b) were carried out in more gas permeable conditions of brown coal beds and younger formations of hard coals. Generally, these deposits occurred at shallower depths, down to 300 m, and ignited relatively easily. Strongly swelling and coking coals have the tendency to block gas flow through the coal bed, thus hindering the course of the reaction. Anthracites have low permeability, they usually occur at great depths, and are non-flammable; however, they are also suitable for gasification.
2) *There is a tendency to consider UCG in deeper beds. This is connected with the simultaneous increase in the unit cost of derived gas.* Until now, practical realisation of UCG was accomplished mainly in coal beds that occur at depths from 300 to 700 m. Trials at considerable depths, though not numerous, took place mainly in Europe, including Belgium (Kowol 1997; Creedy et al. 2001). Analysis of the literature suggests that coal gasification in areas not disturbed by mining activities at depths exceeding 700 m will provide a tight underground generator owing to the plastic behaviour of carboniferous shales in these conditions (Jaskulski and Rabsztyn 1982; Creedy et al. 2001). Conditions are different in areas disturbed by mining activities where the implementation of UCG requires many preparatory operations, including the sealing of gas generators as is the case for the Upper Silesian Coal Basin.
3) *For UCG technology, highly inclined coal beds do not impose additional problems.* The gasification of steeply occurring beds will be characterised by fewer drilling operations compared to those for horizontal beds.
4) *The gasification of beds 1 m thick or more improves economics* (Burton et al. 2007; Jaskulski and Rabsztyn 1982). Beds that are thinner than 0.5 m. are not considered suitable for UCG.
5) *European and Polish investigations suggest that beds with ash content exceeding 50% by weight decreases the calorific value of the gas and hinders the process control* (Jaskulski and Rabsztyn 1982). In case of moderate ash quantities in coal, however, the ash oxides of some metals influence catalytically the course of the gasification reaction and increase the calorific value of the produced gas.
6) *The UCG process may be influenced by the presence of water.* The natural water inflow has a considerably greater impact on UCG than the bed moisture itself. The occurrence of aquiferous zones in the direct vicinity of the gasified bed or a strong natural water inflow can cause great difficulties, and even the stoppage of the gasification process. The UCG process can also exert a negative impact on the water level and purity. The maximum content of water for the production of usable low-calorific gas amounts to somewhat more than 0.5 kg of water per 1 kg of coal designed for gasification. Thus, excess water must be pumped out of the system.

HARD COAL RESOURCES IN POLAND

The potential hard coal reserves in Poland concern developed resources in operating mines as well as coal resources and beds that are located in regions with no mining activities, in regions of planned mining activities, as well as coal beds in the regions of existing mines which are not extracted (Tables 2–4). Hard coal deposits occur in the Upper and Lower Silesian and Lublin Coal Basins. In 2000, however, mining operations in the Lower Silesian Coal Basin were terminated. Also, in the Lublin Coal Basin, only one hard coal mine is operational, the "Bogdanka" colliery. Its production in 2003 amounted to approximately 4 million metric tons (Fig. 1). The richest coal seams occur in the Upper Silesian sandstone series, particularly on the main saddle, within the Bytom trough and on the Jastrzębie saddle. The average abundance of coal in developed mining areas amounts to 20–40 mt/m^2. In the deposits of the Upper Silesian Coal Basin, steam coals constitute 2/3 of the total intrinsically economic reserves, while much of the balances are coking coals.

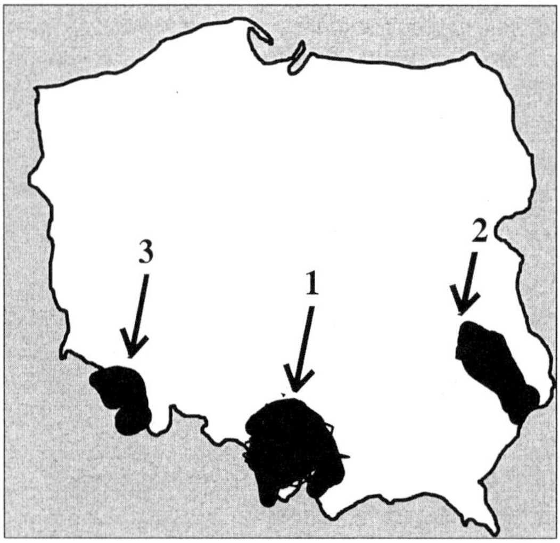

Figure 1. Coal Basins in Poland: 1 – Upper Silesian Coal Basin,
2 – Lublin Coal Basin, 3 – Lower Silesian Coal Basin

Table 2. Hard coal resources in Poland in million Mg

Years	Geological resources	Intrinsically economic resources	Potentially economic resources	Economic resources
2001	75,865	45,900	29,965	7,503
2002	73,566	44,084	29,482	7,381
2003	70,799	43,122	27,677	7,102
2004	69,949	42,579	27,370	6,928
2005	70,592	43,321	27,271	6,014

Table 3. Hard coal resources in Poland in developed and undeveloped deposits in the aspect of mining

Years	Deposits	Resources, million Mg			
		geological	intrinsically economic	potentially economic	economic
2001	developed	29,184	16,045	13,139	7,496
	undeveloped	46,681	29,855	16,826	7
2002	developed	28,868	15,888	12,980	7,376
	undeveloped	44,698	28,197	16,503	4
2003	developed	27,402	15,971	11,431	7,088
	undeveloped	43,397	27,151	16,246	14
2004	developed	35,842	16,041	19,801	6,928
	undeveloped	43,107	26,538	16,569	0
2005	developed	26,682	15,921	10,760	6,012
	undeveloped	44,541	28,030	16,511	2

Table 4. Reserves in developed and undeveloped deposits in Poland as on 31st December 2005

No.	Area	Resources, million Mg		
		Intrinsically economic resources	Potentially economic resources	Economic resources
1.	Poland	15,713	11,143	6,001
2.	USCB (GZW)	15,123	10,716	5,681
3.	LCB (LZW)	590	427	320
4.	LSCB (DZW)	–	–	–

USCB (GZW) – Upper Silesian Coal Basin; LCB (LZW) – Lublin Coal Basin;
LSCB (DZW) – Lower Silesian Coal Basin

The geological reserves in each deposit, determined on the basis of geological documentation, are divided into intrinsically economic reserves and potentially economic reserves (Tables 3 and 4). Taking into account the calculation of economic effectiveness of mining and industry-related criteria determined for every deposit, the economic resources and non-economic reserves are differentiated from the intrinsically economic reserves at the stage of designing the deposit development. Economic hard coal reserves constitute reserves that are classified for mining based on:

- bed thickness;
- depth of bed occurrence;
- quality parameters of coal;
- geological and mining conditions of occurrence in the deposit;
- technical possibilities of mines.

Resources that have not been extracted represent the difference between economic reserves and reserves that will remain in the deposit following mining activities; these are the so-called mining and non-mining losses. The losses are inevitable and result from the limitations of current technical mining methods as well as geological and mining conditions (Collective work 2002).

In Poland, in 2005 there were documented 70,592 million Mg of geological resources of hard coal (Tab. 2). Only part of these resources meets the cut-off parameters, introduced by the regulations of the Geological and Mining Law of 1994. The cut-off parameters used as thresholds are the depth of the seams, the coal thickness and seam thickness, the proportion of the barren partings in the seam, the ash content in the seam and cleaned coal, the calorific value of the raw and cleaned coal, and the sulphur content. The resources that fulfil the cut-off parameters comprise 128 deposits and amount to 43,321 million Mg, of which nearly 1/3 constitutes coking coal resources. The remaining 27,271 million Mg comprise potentially economic resources, i.e., resources which, in the current technical and economic conditions of the country, cannot be the subject of mining activities. In this group are included potentially economic resources that meet the cut-off parameters of coal beds but which cannot now be mined on account of difficult geological and mining conditions, or protection of surface and underground objects. However, it is expected that according to the regulations of the geological law (Geological and Mining Law 1994) they can be mined in the future as a result of technical progress, economic, organizational and administrative changes, or investments.

The analysis of the mean of coefficients relating to the use of resources, which for balance reserves amounts to 0.47 and for potentially economic reserves to 0.39, indicates a high potential of coal resources that can be considered as the basis for coal gasification projects.

Mineable resources of about 6,004 million Mg of coal represent resources that can be extracted; their final quantity will be the difference between ascertained commercial resources and resources which will remain in the deposit because of the lack of extraction possibilities, i.e., the losses in deposits. The losses make up on average about 20% of economic resources. These losses remain in the deposit as a potential resource for the future.

It is also of interest to consider the resources suitable for new techniques of energy production, including UCG. Apart from resources developed for coal mining in operating mines, in the total quantity of coal resources in Poland there exist undeveloped geological resources (Tab. 3), composed of:

- the resources of prospective deposits;
- the resources of reserve deposits;
- the resources of deposits where mining has been ceased because of mine closure.

Undeveloped resources as well as resources remaining in closed mines can also constitute the basis for underground energy generation processes from hard coal. Thus, Polish coal resources for UCG may be considered to be among the largest in the Europe.

Generally the reserves of Polish coals are characterised by good quality parameters. About 73% of them constitute coals with calorific value exceeding 25 MJ/kg. A very small quantity amounting to 6% points out a calorific value below 22 MJ/kg. Almost all reserves mentioned above, i.e. about 92%, include ash amounting to 20%, of which 50% constitute coals with ash content not exceeding 10%. Most of these reserves, as many as 77.4%, have low sulphur content not exceeding 0.9%. The reserves containing more than 1.5% constitute only 5% of the general reserve quantity and occur mainly in mines that extract seams of the highest part of the Carboniferous profile.

Because of the seam inclination angle, the reserves are relatively easy for mining operations, because they occur mainly in seams with low inclination. 76.7% of reserves occur in seams with inclination up to 12°, 22% in seams with the angle of inclination from 13 to 30°, and only 1.3% of reserves in seams, where the angle of inclination is higher that 30°.

About 53% of coal reserves occur in seams of medium thickness, i.e. from 1.5 to 3.5 m. Only 25% occur in seams of thickness exceeding 3.5 m, and 22% in thin seams of thickness below 1.5 m.

Power coals constitute about 64% of reserves, high-rank coking coals about 24%, while coking coals constitute only about 12%. High-rank coals occur generally in mines situated in the south-western part of the Upper Silesian Coal Basin.

The analysis of hard coal reserves according to such parameters as seam thickness or the angle of inclination of seams as well as quality parameters allows to state that in the majority those are reserves occurring in seams of medium thickness and small inclination, characterised by coals with very advantageous quality parameters, i.e. low ash and sulphur content and high calorific value.

The carried out analysis allowed to assess and select mines in the Upper Silesian Coal Basin (Kaziuk 2001) that have the most convenient conditions for coal gasification, taking into account the tectonics, methane hazards, seismic hazards, fire hazards and rockburst hazards (Fig. 2).

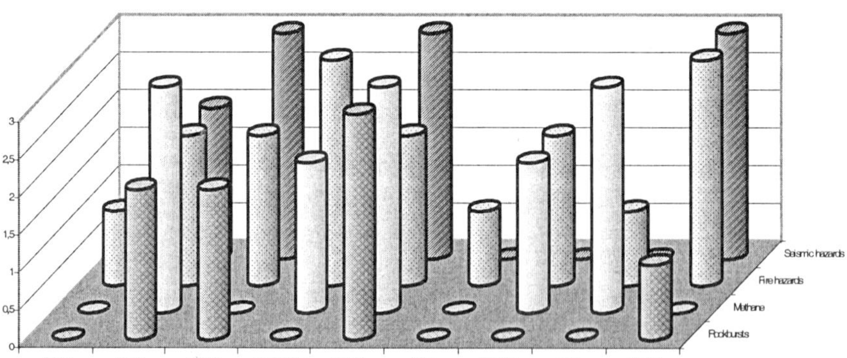

Figure 2. Rockburst, methane, fire and seismic hazards for selected mines
hazard scale: 0 – lack, 1 – low, 2 – medium, 3 – high

LIGNITE RESOURCES IN POLAND

Lignite constitutes a stable and secure source of energy. Due to the fact that its deposits will allow another 300 years of exploitation at the current rates, rapid depletion will not constitute a threat in the most foreseeable future. In Poland, there are 190 identified deposits and lignite bearing areas. The geologically documented intrinsically economic resources total approximately 14 billion Mg, whereas the potentially economic ones are estimated at 4.9 billion Mg. Therefore, they will definitely contribute to Poland's energy security. In light of the above considerations, it is of key importance to carry out an analysis on their potential utilization in the process of underground gasification as such a mode of extraction allows the protection of landscapes and the natural environment conservation. The advantages of the conversion of lignite into gas directly in the seam are twofold; firstly, the necessity to relocate huge masses of overburden, which leads to often irreversible devastation of the landscape, is eliminated and secondly, serious ingerention into underground water regimes is not required. The technology of underground coal gasification may be applied in cases where the exploitation by means of open cast mining is impossible or uneconomical.

Lignite deposits in Poland were formed in the peripheral zone of an extended Tertiary basin of North-Eastern Europe and constitute the continuation of the coal seams exploited in the Lusatioan region, Germany. Seams of particular thickness were formed in tectonic foredeeps (the deposits of Bełchatów, Czempin, Gostyń, Krzywiń, Lubstów, Mosina, Naramowice, Radomierzyce, Szamotuły, Turów) or in the foredeeps located above salt domes (the deposit of Rogóźno). The formation of numerous smaller deposits is associated with the upheaval of Miocene sediments. Such smaller deposits are exploited only in the regions of Bełchatów, Konin-Turek and Bogatynia (Piwocki et al. 2004). Figure 3 illustrates the distribution of lignite deposits in Poland; Table 5 reviews lignite resources in Poland for the period 2001–2005.

Figure 3. Distribution of lignite deposits in Poland (Piwocki et al. 2004)

Table 5. Lignite resources in Poland in million Mg (PIG 2005)

Years	Geological resources	Intrinsically economic resources	Potentially economic resources	Economic resources
2001	18,800	13,924	4,876	1,757
2002	18,543	13,860	4,683	1,697
2003	18,246	13,685	4,611	1,635
2004	18,251	13,635	4,616	1,586
2005	18,316	13,724	4,592	1,527

Table 6 presents the characteristics of lignite resources in the aspect of developed and undeveloped deposits for years 2001–2005.

Table 6. Lignite resources in Poland in developed and undeveloped deposits in the aspect of mining

Years	Deposits	Resources, million Mg			
		geological	intrinsically economic	potentially economic	economic
2001	developed	2,286	2,077	209	1,722
	undeveloped	16,516	11,847	4,669	35
2002	developed	2,215	2,014	201	1,661
	undeveloped	16,528	11,846	4,682	36
2003	developed	1,968	1,839	129	1,600
	undeveloped	16,328	11,846	4,482	35
2004	developed	1,922	1,789	133	1,551
	undeveloped	16,329	11,846	4,483	35
2005	developed	1,987	1,878	109	1,490
	undeveloped	16 329	11,846	4,483	37

The lignite deposits are mined in Central and Western Poland in four open-cast mines ("Sieniawa", the fifth one operates only to satisfy local needs), three of which are multi-pit open cast mines ("Adamów", "Bełchatów", "Konin"), and the remaining one ("Turów"), is a single-pit mine.

The most abundant deposits of lignite occur in Bełchatów at a considerable depth with the overburden thickness of 150 m; nevertheless, the thickness of the very seam which equals approximately 50 m makes its exploitation very economical. In terms of lignite mining, the exploitation of a deposit is regarded as economical when the ratio of overburden thickness to coal thickness is 6 to 1. In case of Bełchatów, the ratio is 3 to 1 enabling particularly favourable open cast mining exploitation of the deposit. The deposit of Turów in the region of Bogatynia, as well as the Konin deposits (Pątnów, Lubstów and Adamów) are characterized by much less advantageous geological conditions than those in Bełchatów, although the depth of bed occurrence is smaller. It is also worth mentioning that the Szczerców field within the Bełchatów deposit is currently under development.

The total lignite output in 2005 equaled 61,612 mln Mg remaining at same level as in 2004. The intrinsically economic resources of Polish lignite amount to 13 724 mln Mg, of which bituminous coal accounts for 0.8 mln Mg, briquette coal accounts for 2515 mln Mg, whereas carbonized coal – 1458 mln Mg. However the total amount is utilized and regarded as energy coal. Since the beginning of lignite mining operations in Poland, over 2 billion of these energy carriers were exploit-ted, which also required the removal of 8 billion m^3 of overburden. The exploitation of the deposits is currently carried out by means of open-cast mining and the lignite is combusted in nearby power plants. Underground gasification may constitute an alternative to the current methods of production.

The chemical and technological parameters of lignite result from its position within the carbonization process, the parameters are placed between those characteristic of peat and hard coal. Typically, lignite is characterized by low calorific value, low consistency, as well as high water and ash content, which contribute to problems associated with its transport. For Polish soft lignites which almost entirely constitute the country's lignite resources, the parameters occur in the following ranges:

1. calorific value Q^r 8,0–9,3 MJ/Mg;
2. ash content A^d 18–27%;
3. total Sulphur content S_c on average 1,6%;
4. natural moisture W^r on average 53%.

The above data illustrate that Polish lignite represents good quality and therefore it may be utilized for energy purposes.

While analyzing the resources of lignite, special consideration should be taken to the undeveloped deposits which have not been subject to mining operations so far. Underground coal gasification may be applied for the seams that cannot be exploited by means of classic open cast mining methods due to technical or economic restraints. Table 7 below reviews basic parameters of the main deposits which are currently undeveloped (75 mln Mg).

Table 7. The characteristics of main undeveloped deposits

Deposit	Thickness, m	Depth, m	Calorific value, kcal/kg	Ash content, %	Average sulphur content, %
Babina-Żarki	10,7	140,0	2 229	18,28	1,10
Cybinka	16,6	94,0	2 236	18,40	1,41
Gubin	10,2	82,0	2 240	12,86	1,42
Głowaczów	4,8	37,1	1 820	28,56	0,42
Legnica p. Północ	23,0	193,2	2 164	18,94	1,58
Legnica p. Wschód	19,8	136,3	2 206	19,05	1,33
Legnica p. Zachód	21,0	158,8	2 371	20,10	0,76
Mosty	9,3	105,0	2 219	17,19	1,63
Piaski	6,1	48,5	2 082	24,80	1,44
Rogoźno	35,0	195,0	2 319	18,90	3,99
Rzepin	12,2	97,3	2 164	15,14	1,20
Sądów	12,2	127,5	2 196	18,80	1,38
Ścinawa	21,4	213,4	2 390	10,69	0,48
Torzym	21,4	180,8	2 270	16,80	1,81
Trzcianka	4,6	46,4	2072	19,46	1,84
Złoczew	46,2	259,1	2 021	21,67	1,18

Taking into consideration bed thicknesses, coal quality parameters, depths of bed occurrence, geological and mining conditions, as well as the fact that the deposits have not been subject to mining operations, of particular UCG interest are the ones located in the prospective basin of Legnica and Głogów. Those deposits are under protection as they are of strategic importance; thus the region is not considered for housing and infrastructure development.

THE REQUIREMENTS OF A UCG SITE

UCG may safely operate under many different conditions and ranks of coal. However, in the near term, environmental concerns prompt consideration of site characteristics well before burn initiation:

- potential sites must meet minimal requirements (Tab. 8);
- stratigraphic and structural characterization is needed to satisfy information requirements for risk protocols;
- preferred consideration should be given for sites deeper than 200 m;
- preferred consideration should be given for sites with strong or rigid overlying strata;
- downgraded consideration should be given to sites where the coal seams or surrounding strata act as aquifers;
- downgraded consideration should be given to heavily deformed structures or steeply dipping seams.

Table 8. Minimal requirements for UCG sitting and operation

Criterion	Characteristics/remarks
Planning issues	– proximity of neighbours (limited human activities in vicinity); – recommended lack of building development; – environmental sensitivity
Geology	– coal seam > 1,5 m thick; – depth between 150 and 1200 m; – the availability of good density borehole data
Hydrogeology	– positions of aquifers (>100 m vertical separation from major aquifers); – inflow risk and flow potential
Mining	– distance from any recorded and possible unrecorded workings (stand-off of > 500 m from abandoned mine working, licensed areas); – not marked for conventional mining; – not adjacent to working mines
Gas use	proximity to small-scale industrial user or suitable electrical grid connection
Social perception	environmental impacts are acceptable

The worldwide UCG experience demonstrates that avoidance of environmental contamination in future operations can be achieved but should involve integration of criteria for site selection with choices of operating parameters.

Some of the steps that can be taken to avoid the situations that caused past groundwater pollution problems include:

- balancing operating conditions to minimize outward transport of contamination from greatly overpressurized burn zones;
- ensuring UCG sites are situated where geologic seals sufficiently isolate the burn zone from surrounding strata;
- selecting sites with favourable hydrogeology to minimize widespread movement of the contaminated groundwater plume;
- isolating UCG locations from current or future groundwater resources.

A preferred UCG site would involve development of a gasifier in coal measure strata. The target seam should ideally consist of at least 2 m of non friable, fault-free coal with a weak roof to encourage caving and possibly a relatively strong floor to assist drilling guidance.

No significant aquifers should be present within the expected zone of disturbance. There should be no mining activities, current or planned, near the site.

All sets of criteria for a satisfactory UCG area are based around:

- Establishing that it is an economic resource of suitable size.
- Geological conditions are suitable for consistent coal removal.
- Environmental impacts are acceptable.
- A comprehensive analysis will still have to be performed to ensure that the site is suitable, but use of simple criteria can eliminate unsuitable sites quickly.

The criteria for selecting UCG sites in Poland have been formulated in this paper. Based on the available information on the characteristics of coal seams in Poland, it is recommended to focus on the south-western part of the Upper Silesia Coal Basin and prospective basin of lignite of Legnica Region.

CONCLUSIONS

The analysis of the hard coal resources in mines of the Upper Silesian Coal Basin indicates the possibility to use part of its reserves for UCG.

- In the structure of geological resource of all the analyzed mines, there exist currently economically unrecoverable deposits. It is expected, however, that they could be mined in the future. The quantity of these resources as of 31st December, 2005 amounted to *27,271 million Mg* on a country scale, of which *10,760 million Mg* occur in developed deposits.
- Recoverable resources constitute only a part of the coal resource basis in Poland. These resources are determined on the basis of the criteria of the current potential of their extraction, i.e., within the period of the next several years. Nevertheless, their quantity may be subject to change. Both losses and considerable increases resulting from changes in the deposit can occur in them. Thus, the quantity of these resources should be considered only as a rough estimate in case of energy generation techniques other than the traditional ones.
- The losses in coal resources resulting from mining activities are high. They amount on the average to about 20% of commercial resources. These remaining commercial resources which, due to geological and mining conditions of the deposit were not mined, can be considered for UCG processes.
- The analysis of the mean of coefficients relating to the use of resources, which for balance reserves amounts to 0.47 and for potentially economic reserves to 0.39, indicates a high potential of coal resources that can be considered as the basis for coal gasification projects.
- The most convenient parameters with respect to the realization of underground coal gasification, taking into consideration the tectonics, methane hazards, seismic hazards, fire hazards and rockburst hazards, were found for the following mines: Janina, Murcki, Jankowice, Ziemowit and Chwałowice.
- The potentially economic resources, losses and reserves of deposits where mining activities were ceased because of mine closure (Pniówek, Borynia, Krupiński, Pokój, Katowice-Kleofas, Kazimierz-Juliusz, Wieczorek, Polska-Wirek, Piast, Piekary), form the basis for non-traditional energy generation technologies.

The analysis of Poland's resources of lignite leads to the following conclusions:

- Low rank lignite occurring in numerous locations in Poland may be considered for UCG due to its high reactivity, moisture content and small depth of occurrence.
- From the perspective of environmental protection, it could be beneficial to put under discussion the idea of utilizing the undeveloped and unmined lignite deposits by means of UCG.

- The most convenient parameters with respect to underground coal gasification, i.e., the bed thicknesses, depths of bed occurrence, coal quality parameters, were found for the prospective basin of Legnica and Głogów.
- Mean coefficients of utilizing the resources of lignite, 0.14 for intrinsically economic resources and 0.024 for potentially economic ones; indicate considerable potential to utilize them as a basis for future UCG projects.

It should be stressed that the analysis carried out on the basis of mineable resources does not reflect the total quantity of resources available for future development considering different techniques, including gasification technologies.

REFERENCES

Białecka B. 2007a: Analysis of the Impact of Natural and Technological Factors on the Underground Coal Gasification Process. Mining Technology (in press).

Białecka B. 2007b: Estimation of Coal Reserves for UCG in the Upper Silesian Coal Basin, Poland. Natural Resources Research, 4/2008, vol. 17, no. 1.

Bilans Zasobów Kopalin i Wód Podziemnych w Polsce wg stanu na dzień 31.12.2005 r. PIG, Warszawa 2006.

Blinderman M.S. 2005a: The Energy Underground Coal Gasification and its Application in Commercial Clean Coal Projects. Second International Conference on Clean Coal Technologies for our Future, 10–12 May 2005, Cagliari, Italy, http://www.cct2005

Blinderman M.S. 2005b: UCG History. http://www.ergoexergy.com/eUCG_his.htm

Burton E., Friedmann J., Upadhye R. 2007: Best Practices in Underground Coal Gasification. Lawrence Livermore National Laboratory, pp. 52–77. https://eed.llnl.gov/co2/pdf/BestPracticesinUCG-draft.pdf

Collective work (19 experts) 2002: Analysis of Production Possibilities of Polish Hard Coal Mines, Taking into Account their Resources Basis and Work Safety, Environmental, Social, Technical and Economic Aspects. (In polish), p. 89. http://www.mg.gov.pl/NR/rdonlyres/2913591E-B88A-453C-B4CC-233510A37634/0/. Raport na temat reformy górnictwa węgla kamiennego w Polsce 2003/2006.doc

Creedy D.P., Garner K., Holloway S., Jones N., Ren T.X. 2001: Review of UCG: Technological Advances Report No. COAL R211. Aug. 2001, Crown Copyright, p. 36, available on DTI web site, http://www.berr.gov.uk/files/file18660.pdf

Creedy D.P., Garner K. 2004: Clean Energy from Underground Coal Gasification in China. DTI Cleaner Coal Technology Transfer Programme, Report No. COAL R250 DTI/Pub URN 03/1611, pp. 15–17.

Darski J., Kicki J., Sobczyk E.J. 2001: Report on the State of Economy of Hard Coal Deposit Resources. (In polish), IGSMiE PAN (Mineral and Energy Economy Institute of the Polish Academy of Sciences), pp. 22–26.

Geological and Mining Law. (In polish), Law of 4 February 1994. http://isip.sejm.gov.pl/servlet/Search?todo=open&id=WDU19940270096

Hankus A., Białecka B. 2005: Balance of Domestic Resources for Underground Coal Gasification. (In polish), Scientific Publications of the Central Mining Institute, Mining and Environment, No. 4, pp. 67–77.

Jaskulski Z., Rabsztyn Z. 1982: The Role of Natural Factors in the Process of Underground Coal Gasification. (In Polish), Wiadomości Górnicze, No. 2, pp. 45–48.

Kaziuk H. et al. 2001: Report on the Statutory Activities of the Central Mining Institute entitled "Bank on Geological Coal Resources". (In polish), Central Mining Institute Press.

Kowol K. 1997: Chances and Perspectives of Underground Coal Gasification. (In polish), Proceedings of the School of Underground Mining, IGSMiE PAN (Mineral and Energy Economy Institute of the Polish Academy of Sciences), pp. 119–126.

PARGWK S.A. 2005: Resources Balance. PARGWK Press (PARGWK S.A. – State Agency of Hard Coal Mining Restructuring, Joint-Stock Company) (in polish), pp. 11–47.

Piwocki M., Kasiński J.R., Saternus A., Dyląg J.K., Gientka M., Walentek I. 2004: Aktualizacja bazy zasobów złóż węgla brunatnego w Polsce. Warszawa, PIG, Centr. Arch. Geol.

Rauk J. 1979: The State of Coal Gasification in the World. (In polish), Scientific Publications of the Central Mining Institute, Central Mining Institute Press, pp. 1–9.

International Mining Forum 2009, Sobczyk, Kicki & Saługa (eds) © 2009 Taylor & Francis Group, London, UK, ISBN 978-0-415-80428-8

Synthetic Stowing of Rockmass at Borehole Underground Coal Gasification (BUCG)

Volodymyr I. Bondarenko, Volodymyr S. Falshtynskiy, Roman O. Dychkovskiy
National Mining University, Dnipropetrovs'k, Ukraine

ABSTRACT: In the paper it is considered the mechanism of cavities forming in the rockmass. The geometrical parameters of their distribution are calculated. The parameters of the goaf injection stowing under gas generator are defined. Economical values of gas generator usage are presented.

KEYWORDS: Underground coal gasification, technological parameters, components of pressure

World energy realities result in to strain of relations in the energy market. Deficit of reserves of oil and gas increase the coal price. Such situation is important for Ukraine as gradual exhaustion of reserves results in necessities of mining of non-commercial coal reserves and magnification of competitiveness of the Ukrainian fuel.

Borehole underground coal gasification (BUCG) is one of quantity new mining technologies. But a number of the deficiencies related to hermetic stowing of an underground gas generator do not allow supervising over gasification process effectively.

Tightness pinch, and thereof – reliabilities of an underground gas generator by injection stowing material in the distorted soils of roofing and goaf is solution of these problems. Lack of sufficient substantiations of technological parameters and key diagrammes of putting of goaf at UCG provides analysis of dynamics of the temperature profile, the intense-deformed space rock-coal massif and mining of parametres of technology of a goaf stowing taking into account change of material and the heat balance of process of gasification of a coal seam in time space.

According to the analysis of the mechanism rock mass reversion at gasification of coal NMU method is chosen, it allows to develop mathematical model of reversion of a rock mass taking into account features of process of gasification of coal seams. As data for calculation the mining-and-geological performance of the chosen site and technological parameters of process of gasification served stratigraphic sections on boreholes. According to measure of operability of coal seams to gasification in place sites of a carboniferous deposit of the Western Donbass SC "Pavlogradvugillya" with stratification depth of 170–340 m have been chosen.

Concrete geological conditions of a site of gasification and technological parameters of process of BUCG during calculation are entered in SC. Effects of calculation geometrical, physical properties diagram loadings, lowerings and horizontal travels of rocky seams of the moved massif in the form of tables, graphs and diagrams are displayed on the monitor.

The executed examinations of the intense-deformed state of rock stratum with a substantiation of dynamics and parameters formation of cavities an unmixing of a laminated massif which contains a gas generator, have allowed to position presence of cavities an unmixing and changes of their sizes depending on natural factors and technological parameters.

At gasification of a coal seam on the chosen sites of coal deposits over goaf apart 9,5–14,6 m are shaped unmixing cavities.

Changes of voidage of an unmixing, taking into account change of technological parameters (length of the hot-fire channel of an underground gas generator and velocity of gasification of the coal seam), occurs on a linear relation.

To the extent of a motion of hot-fire working face and magnification of goaf the unmixing cavity is closed, and after the termination of operations on gasification of a coal seam at length of a pole which outgases, 100 m, its height do not exceed 76 mm, and volume 81 m^3 depending on change of technological parameters. The cavity volume changes from 610 to 924 m^3, the unmixing height fluctuates in boundaries of 181–326 mm.

Reversion of soils of a main roof of an underground gas generator is not under the influence of thermodynamic loadings (because of small heat diffusivity and major heat capacity of rocky seams) an immediate roof. Under the yielded requirements the smooth lowering of a main roof with formation of vertical fractures and unmixings will be observed.

The immediate roof at gasification of a coal seam under the influence of forces of gravitation and loadings exfoliates. Such state of roofing will depend on parameters of a loading from a main roof, heats, power of ashes, presence in a roof and a sole of the soils predisposed to inflation.

The seam immediate roof, in the core is presented by batts which fall into thermally sound plastic soils. Considering irregularity of allocation of temperatures in the reactionary channel of an underground gas generator, at a maximum of temperatures 1200°C on boundary of transferring of an oxidizing band in reduced (length of transferring 0,74–1,5), the lapse rate of warm-up of soils of roofing will make 30,1–52,4°C/0,1 m.

At warming-up of soils of roofing to 1000°C on depth of 0,45 m from hot-fire working face the unmixing and fracturing of integrity which is signs of inflation [1] is observed. In temperature range of warm-up 950–1200°C on depth of 0,1–0,35 m intensive inflation of soil and transferring of soils in plastic condition (melting) longwise transferring boundaries is observed.

Analyzing effects of calculations, it is necessary to score, that at magnification of travelling speed of hot-fire working face an abutment band which arises at an accuration of a rocky seam of roofing, decreases in sizes from 11 to 6,8 m.

Changes inappreciably with detrusion more close to working face which result into strains with velocity of 4,8–5,7 mm/minutes At such strains in a rocky seam the active formation of horizontal and vertical fracturing is observed.

Soils of an immediate sole of a seam are inclined to inflation on 78–446 mm, taking into account power of ash dross (9,6–10,8 mm) it makes half of power of gasificated seam. As a result of these processes the smooth lowering of a rocky seam of roofing on bellied sole with integrity tearing up, in the form of horizontal and vertical fracturing will be observed.

Examination of dynamics formation and temperature profile reversion in a rock mass at gasification of a coal seam was made on test bed and laboratory installations and analytically.

Test bed and laboratory examinations were made in steady-state conditions on mine № 1 "Ostriy", coal mine "Kurahovs'ke", mine "Ternivs'ka" and in NMU laboratories.

Effects of experiments give the chance to ascertain, that apart 1,5 m (6 m in nature) from a coal seam, roofing soil give in to intensive warming-up by light-end products BUCG, at the expense of their migration on fractures and unmixings in a percarbonic massif with 1,5 to 2,3 m (6–9,2 m in nature). Intensity of warm-up falls at the expense of reduction of the sizes of failures in stratum of soils.

Apart 0,4–1,6 m from a coal seam temperature of warm-up of soils change from 910 to 400°C. Losses of temperatures on this site make 46%. Apart 0,4–0,9 m (in nature 1,6–3,2 m) temperature fluctuate from 420 to 111°C, losses of temperatures in a massif on this site are equalled 26,4%. On

54°C warmed-up soils of roofing to 6 m. Then temperature apart 2,25 m (in nature 9,0 m) from goaf come nearer to a natural lapse rate 25–20°C. At magnification of pressure to 0,6 MPa and application oxygen blow intermixtures heat penetration of a rock mass to 50°C will make 6,4 m that is related to strain of soils of roofing and magnification of velocity of migration of high-temperature light-end products in rock massif. It is necessary to score, that the degree of warm-up of soils of a percarbonic stratum depends on pressure in a gas generator, content of blow intermixtures, degrees of an unmixing and fracturing of soils of roofing.

At analytical solution of heat exchange of yields of gasification with soils the quasi-stationary mode, no isothermal fluidity of high-temperature yields taking into account their heat exchange with rocks was considered at BUCG. The rock massifs of soils are taken over isotropic with some centre heat conductivity. On a soil surface there is a heat exchange with light-end products which aspire under the influence of the pressure created in an underground gas generator, to make through deep into a percarbonic stratum on vertical fractures and unmixings which arise under the influence of a rock pressure at gasification a coal seam. Heat emission coefficient in these requirements is so huge, that we will admit to take over requirements of 1st sort on boundary gas BUCG – soil.

As a result of examinations it is positioned, that the peak boundary of a temperature profile located over the hot-fire channel on boundary of transferring of an oxidising band in regenerative with temperature 950–1200°C, breadth of a band of transferring of 0,74–1,5 m, height of distribution of temperatures to 9,0 m from a coal seam to perpendicularly bedding of rocky seams (Fig. 1).

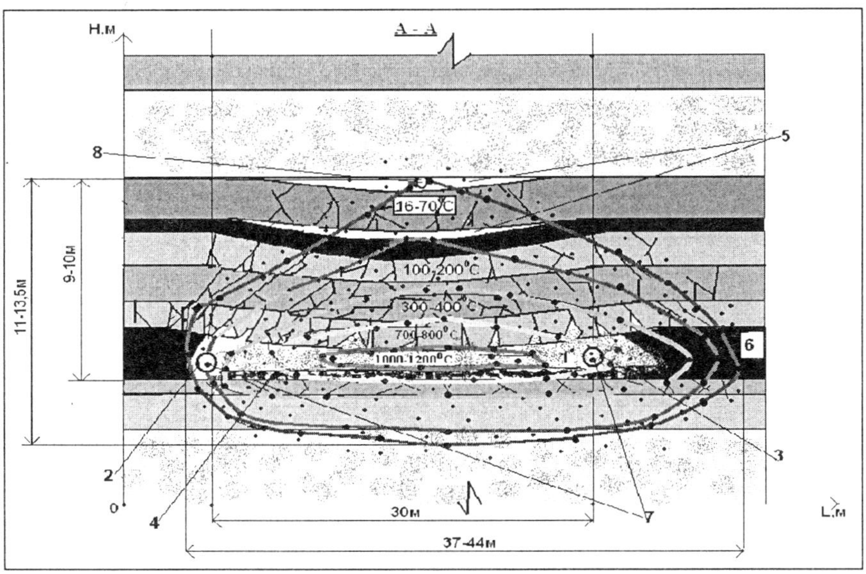

Figure 1. Warming-up parameters of rock-coal massif across to a bedding: 1 – the reactionary channel of an underground gas generator; 2 – blow inclined-horizontal borehole; 3 – an exhaust inclined-horizontal borehole; 4 – ash drosses; 5 – cavity of an unmixing of soils of roofing; 6 – a coal seam; 7 – a hose line; 8 – a stowing pipe

On the basis of the spent examinations and a practical operational experience of stations "Pidzemgas" three constructions of an underground gas generator with different flow diagrammes of disclosing and conducting process of injecting-stowing operations are offered at gasification of a coal seam [2].

As stowing material it is supposed to use muds on a basis of kaolin and bentonitic clays. Clay injective intermixture under the influence of temperatures sluggishly loses the transportable properties and, swelling up, expands, isolating pores, fractures and an unmixing in a rock mass. Expenses of injective solution, depending on travelling speed and length of the hot-fire channel of an underground gas generator, make 95,8–478,3 t/days.

On the basis of the spent examinations of geomechanical parametres of reversion of rock massif and the temperature factor atgasification of thin coal seams field parametres of injecting of soils of roofing and goaf are proved taking into account time and space (tab. 1).

Table 1. Injecting of a percarbonic stratum of an underground gas generator

№	Injection time, hour		Average penetration, D	Average gleam of fractures, m	R_X / R_Y	Forcing radius, m	
	T_x	T_y				R_x	R_y
1.	0,5	0,5	894	$4{,}86 \cdot 10^{-3}$	2,48	7,2	2,9
2.	1,0	1,0			2,97	13,4	4,5
3.	1,5	1,5			3,38	18,3	5,4
4.	2,0	2,0			3,44	22,7	6,6

Field parameters of rock stratum injection which contains a gas generator, depend on fracturing of soils, time of injection, pressures, the sizes of cavity in a massif and warming-up of soils of roofing (Fig. 2).

Physical and geometrical parameters of the yielded field are the following. The inferior boundary of stowing (Ry = 6,6 m taking into account setting clay of injective material under activity convectional heat fluxes with centre temperature T = 317,8°C); filling high bound (Ry = 1,5–2,4 m, change depending on stowing method (self-exact or air-operated); formation field of injective-stowing massif with the active area S = 149,8 m^2; boundaries of the active distribution of a filling with under load activity of temperatures T = 172,4–65,8°C. The band of the upper passive boundary of backfill depends on stowing method and a degree of natural and technogenic fracturing of rocky seams behind bedding and is perpendicular to it.

Lowering of a surface in the conditions of application of injective backfills at gasification of thin coal seams does not exceed 12–18% from power of a coal seam [3].

As a result of analytical examinations the received dependences of losses of yields BUCG at change of pressure and tightness of soils of roofing are presented on Figure 3.

At magnification of pressure with 1,0 to 3,0 MPa on hot-fire working face losses blast and gas are incremented and reach 7,0–21,4% at length of the hot-fire channel of 25 m and 7,5–24,8% with magnification of length of the channel up to 30 m. These parameters sharply drop at hermetic stowing of fitchery grounds of roofing of injective-stowing intermixture. So, at change of pressure in a gas generator on 1–3 MPa losses make 2,4–11,3%. Thus, reduction of fracturing of a laminated stratum of soils by an injection in a band of failure of muds will allow reducing migration of high-temperature yields of BUCG at soil which contain an underground gas generator.

For definition of parameters of pollution by yields of thermal decomposition of hills at gasification of coal seams there was an executed calculation of the mass balance of process of BUCG taking into account a yield of chemicals (condensate) with BUCG gases at different constructions of underground gas generators (Tables 2 and 3).

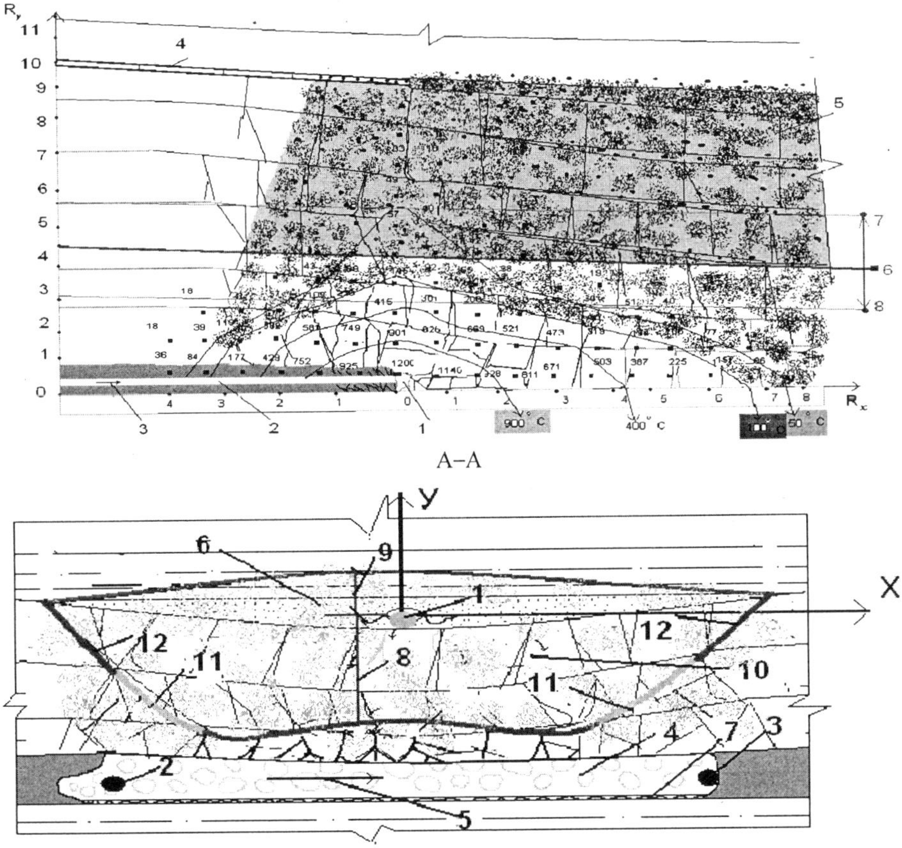

Figure 2. Fields of injection of rock stratum over a gas generator: 1 – the reactionary channel of a gas generator; 2 – a blow borehole; 3 – a motion direction of blow intermixtures; 4 – a stowing borehole; 5 – an injective-stowing massif; 6 – the inferior boundary of injection under the influence of pressure and time; 7 – high bound of temperatures which influences properties of relaxation of injective mud (50–180°C); 8 – the inferior boundary of activity of temperatures 180–400°C on a mud (reduction of transportable ability, inflation and sintering of clay before formation of a conglomerate with the distorted soils of roofing)

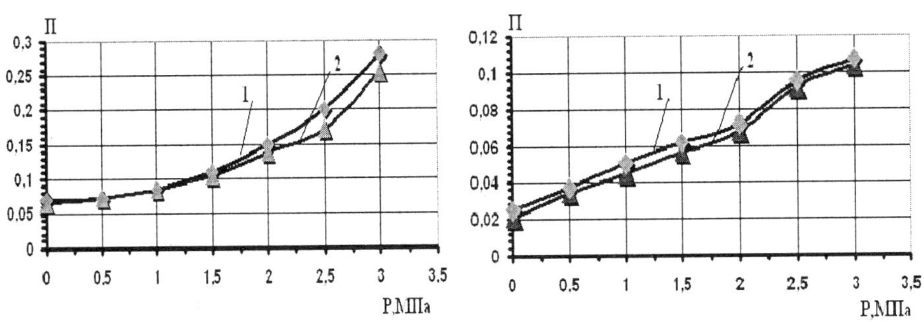

Figure 3. Losses of yields of BUCG depending on pressure in the reactionary channel and tightness of soils of roofing

Table 2. Effects of analytical experience of the mass balance of process of BUCG at different kinds of blasting

Blast intermixtures	Composition of blasting, %			Composition of BUCG gases, %							$Q_{T.gas,}$ MJ/m^3	η_x efficiency factor, %
	O_2	CO_2	vapour	H_2	CH_4	CO	N_2	H_2S	CO_2	O_2		
O_2N_2	23	-	-	9,4	5,7	18,9	55,3	0,35	9,3	1,2	5,4	72,0
	65			16,3	9,8	33,5	20,0	0,8	16,0	3,2	9,6	76,0
	60			17,2	10,7	34,9	20,1	0,7	14,3	2,6	9,8	77,0
$H_2O + O_2$	23	-	8	10,1	6,0	19,3	54,5	0,5	8,0	1,4	5,7	73,0
	55		19,6	17,5	10,5	35,4	6,0	1,0	25,4	4,0	10,2	78,0
	55		34,0	19,2	11,5	28,9	7,2	1,1	18,6	3,3	11,2	82,0
$CO_2 + O_2$	55	41,8	-	17,9	10,1	48,1	1,8	1,2	17,3	34,3	11,6	86,0
	50	32,6		16,9	9,6	43,8	2,1	1,1	22,0	3,4	0,9	85,0
	50	43,5		17,5	9,7	47,8	2,2	1,1	18,0	3,4	11,5	86,0
$CO_2 + O_2 +$ $+ H_2O$	50	32,6	13,3	18,1	10,3	46,9	2,5	0,9	17,7	3,3	11,7	87,0
	45	39,1	12,0	17,7	9,9	48,2	2,41	0,8	17,5	3,2	11,7	87,0

Table 3. The mass balance of chemicals of a condensate at gasification of coal seams of mark DG

The kind of blow intermixtures	Yield of chemicals gasification, kg/day							
	Stown-coal resin		Crude benzol		Amianth		Sulphur	
	Without stowing	With stowing	Without stowing	With stowing	Without stowing	With stowing	Without stowing	With stowing
O_2N_2	670	773	584	673	1569	1806	148	155
H_2O(пар) + O_2	700	805	599	689	1394	1601	135	141
O_2(30–62%)N_2	747	861	624	739	1330	1531	122	125
$CO_2 + O_2$	776	893	706	813	1311	1509	139	145
$CO_2 + O_2 + H_2O$	760	874	726	837	1304	1501	131	138

Proceeding from the received effects, and also on an operational experience of stations "Pidzemgas", it is possible to ascertain, that qualitative and quantity indicators of chemicals of BUCG depend not only on quality of coal, geological conditions, content of blowing intermixtures, but also from application in constructions of underground gas generators of backfill of a percarbonic stratum of soils and goaf [4], [5].

Application at BUCG of underground gas generators with injective backfill to allow to reduce the cost price of the received gas on 33,6–45,1% that is provided at the expense of magnification of the area of the installed gas coal in 2,2 times, degrees of gasification of a coal seam on 7% and pinch of warming-up abilities of gas on 61%.

The certain cost price of producer gas on underground gas generators with injective backfill makes 180–257 hrn*/1000 m^3 gas, the price of the electric power received from this gas makes 0,09–0,26 hrn/1kvt.hour, without profit on embodying of chemical raw materials and reclaimed heat (Tab. 4).

Table 4. Price of 1000 kvt.hour, at application of different coal and producer gas, hrn

Index	Fuel			
	Raw coal	Gas		
		Coal gasification		
		The land	Underground, with gas generator preparation	
			From an earth surface	In mine conditions
Capital investments, including:	339,4	401,2	179,8	256,7
Electric power manufacturing	61,4	28,5	17,5	21,4
Fuel extraction	197,1	205,7	-	-
Gas manufacturing	-	80,5	85,3	94,2
Operational expenses	56,7	41,3	47,8	103,0
The given expenses	44,2	45,2	30,4	38,1
Ratio of expenses, %	100,0	117,2	53,0	60,3

Expenses and the cost price of production produced from gas and the chemical combinations received at thermochemical processing of a coal seam are given in Table 5.

Table 5. Expenses and the cost price of produced production at BUCG (*$1 USA = 7,9 hrn Ukraine)

The name of indexes	Measuring units	Index			
		Energetics	The chemical industry	Fuel	Municipal services
One-time expenses	Thousand hrn.	3087	2796	1564	2770
Annual expenses	Thousand hrn.	457	341	237	316
The cost price of the produced electric power	hrn./kVt.hour	0,09–0,11 0,14–0,21*	-	-	-
The cost price of the produced fuel	hrn./kg	-	-	2,12	-
The cost price of the produced producer gas	hrn/1000 m^3	-	148-172	-	245–275
Annual economic benefit	Thousand hrn.	5954	3060	1809	2918
Time of recovery of outlay	Years	1,6–2,0	1,5	1,35	0,95

**At coal underground gasification in mine conditions

Index of investment appeal of developed radical technology of BUCG at the interest rate to 42%, a coefficient of advantage of injected solutions 1,5–2,2.

CONCLUSION

– Velocity of disclosing of cavities an unmixing in rocky (stratification depth 210–340 m) at gasification of a coal seam on sites, changes with magnification of depth at 6–9%.

- At a variation of length of the hot-fire channel from 25 to 60 m and travelling speeds in boundaries of 0,5–1,5 m/day the peak volume of cavity the unmixing will change in boundaries 610–924 m^3 with the peak height according to 181–325 mm.
- Formation of cavities of an unmixing at examination of sites of coal seams, applicable to gasification, is observed in a rock mass over goaf apart 9,6–14 m on a normal line from an underground gas generator.
- On a gas generator contour over goaf of concealment of cavity of an unmixing there are apart 0,75–2,8 m from regional sites of a seam, which outgasing, with reduction of height of cavity to 76 mm and volume to 81 m^3.
- Above single soil-bridge of cavity an unmixing is not formed, over them there are allocations of normal loadings close to gravitational composite.
- Through presence of an ash content after gasification of a seam and propensity of a sole of a seam to lifting at heats, detrusion of a massif round an underground gas generator decreases for 7–20%.
- Temperature profile parameters in soils which contain a gas generator depend on a geologic structure of a percarbonic stratum and technological parameters of process of gasification.
- Warming-up of soils of roofing of a coal seam, which is installed gas, carried out at the expense of convectional (56,7%) and conductive (43,3 %) heat exchange.
- The high bound of warm-up of soils of roofing on a normal line from a coal seam to natural temperature gradient of a percarbonic stratum makes 8,4–9,2 m and is in a place of transferring of an oxidising band in regenerative, in section over hot-fire working face.
- Warm-up of soils on a bedding in 1,2 times is less, than on a normal line also makes in a massif of 3,9–4,7 m from hot-fire working face. Volume of warmed-up soils in a rock-coal massif to 100°C - 19,6%.
- In goaf caved in soils in the area of a bedding get warm to 50°C from 7,6 to 11,5 m from hot-fire working face. The volume distorted and caved in soils, which are warmed-up to 100°C over goaf makes 35,2%.
- Thermal conductivity of mentioned rock-coal massif, round an underground gas generator, is characterized by convection influence in transmissions of heat and is in cubic dependence on intensity of fracturing of soils, the sizes of fractures and unmixings.
- Application in a construction of an underground gas generator of injective backfills allows to reduce losses of blast, gases and a condensate in an underground part from 4,5 to 17,8%, to reduce surface lowerings over goaf of an underground gas generator by 76–81% from power of a coal seam.
- The new construction of a gas generator will provide conducting process of gasification at high pressure from 0,8 to 5,4 MPa, without tearing up of integrity of rock stratum, at small and centre stratification depths of coal seams which are installed gas (70–350), providing thus a process dirigibility.
- Formation of cavities of an unmixing in a rock mass related to physical-mechanical properties of soils, a structure of a massif which contains a coal seam, and technological parameters of process of gasification in place.
- Hermetic stowing of an underground part of a gas generator allows concentrating sources of yields of BUCG which will provide efficiency "washing away" at after purification of a gas generator after decay of operations on coal gasification.
- Perfection of a construction of an underground gas generator allows to reduce the cost price of the received gas, in comparison with an existing construction, on 45,8%.
- It is provided at the expense of magnification of the area of the installed gas coal in 2,2 times, a degree of gasification of coal on 7–9,2% and pinch of warming abilities of received gases on 61%.

The intrusion of the offered technology at stations "Pidzemgas" will allow to raise mobility of process of gasification in place at the expense of expansion of field of an intrusion; a dirigibility, compactness, ecological compatibility and safety of process, at the expense of reduction of losses; will allow to cure of gasification a coal seam on high pressure and temperatures which will be provided with growth qualitative and quantity indicators of process of gasification.

Decreasing of the cost price of gas on the cores cost articles, such as drilling on 8,6% and gasification on 32% at the expense of use of more progressive constructions of underground gas generators with injective backfill, and also magnification of the area of a coal gasification with one gas generator.

REFERENCES

[1] Savostianov O.V., Falshtynskiy V.S., Dychkovskiy R.O.: Mechanism of Rockmass Control at Borehole Underground Coal Gasification. Scientific Bulletin of NMU, 2007, № 9, pp. 51–61.

[2] Savostianov O.V., Klochkov V.G.: Management the Rockmass. K.: NMKVO, 1992, 276 p.

[3] Parameters of Temperature Distribution in Rocks, Containing Underground Gas-Generator. B.C. Falshtynskiy V.S., Dychkovskiy R.O., Tabachenko M.M., Lapko V.V. Works of International Scientific Conference "School of Underground Mining", D.: NMU, 2007, pp. 225–232.

[4] Bondarenko Volodymyr I., Buzylo Volodymyr I., Falshtynskiy Volodymyr S.: Parameters of Injection Fill above an Underground Gas Generator. Technical, Technological and Economic Aspects of Coal Thin-Seams. Mining International Mining Forum 2007, London–Leiden–New York–Philadelphia–Singapore, A.A. Balkema Publishers, 2007, pp. 89–96.

[5] Falshtynskiy V.S., Dychkovskiy R.O.: Parameters of Injection of Rockmass above an Underground Gas-Ggenerator. NMU, D.: RVK NMU, 2006, № 26, V. 1, pp. 49–56.

International Mining Forum 2009, Sobczyk, Kicki & Saługa (eds) © 2009 Taylor & Francis Group, London, UK, ISBN 978-0-415-80428-8

Decision Making System Based on GIS Technology for Supporting Machinery Maintenance

Justyna Górniak-Zimroz, Radosław Zimroz, Robert Król, Leszek Jurdziak
Institute of Mining Engineering, Wrocław University of Technology, Poland

ABSTRACT: This paper deals with the application of a computer based maintenance system to support belt conveyor maintenance management. Considering the dispersed, wide-area character of the machinery working in the mine, the peculiar nature of the problem was indicated. In order to exploit as much as possible the available data, the GIS environment as the IT platform was selected. In the mining industry GIS software is often used for different purposes (e.g. mine planning). The choice of the database standard based on MS Access was dictated by its simplicity and the widespread use of this environment in the mines. First, a survey of the literature on computer-aided mine engineering is presented. The survey shows a need for a system dedicated to belt conveyor maintenance management problems. Principal assumptions concerning the system were formulated, data sources were defined, methods of analysis were discussed and finally some examples were presented. It is anticipated that the target system will be a "federation" of subsystems. From the proposed data classification it follows that some of the subsystems can be quickly developed because they contain only technical documentation data (cataloguing only). The other parts require additional monitoring systems, tools and procedures. However, the main task is to build procedures which take into account all informative data in order to get specific information/decision/scenarios of degradation. The universality and open character of the GIS platform opens up prospects for fast data integration and the use of such procedures.

KEYWORDS: Belt conveyor, operation, management, GIS

1. INTRODUCTION

This paper deals with the application of a computer based maintenance system to support belt conveyor maintenance management. Considering the dispersed, wide-area character of the machinery working in the mine, the peculiar nature of the problem is indicated. In order to exploit as much as possible the available data, the GIS environment as the IT platform was selected. In the mining industry GIS software is often used for different purposes (e.g. mine planning). The choice of the database standard based on MS Access was dictated by its simplicity and the widespread use of the environment in the mines. First, a survey of the literature on computer-aided mine engineering is presented. The survey shows a need for a system dedicated to belt conveyor maintenance management problems. Principal assumptions concerning the system are formulated, data sources are defined, methods of analysis are discussed and finally some examples are presented. It is anticipated that the target system will be a "federation" of subsystems. From the proposed data classification it follows that some of the subsystems can be quickly developed because they contain only technical documentation data (cataloguing only). The other parts require additional monitoring systems, tools and procedures. However, the main task is to build procedures which take into account all informative data in order to get specific information/decision/scenarios of degradation. The universality and open character of the GIS platform opens up prospects for fast data integration and the use of such procedures.

2. LITERATURE SURVEY

In recent years attempts have been made by the mining companies to apply IT tools to mine management. Unfortunately, most of the tools are merely programs aiding in the planning of surveying, geologic and mining work. Even though some of them include data on selected mining machines and equipment, they lack a module which would support the management of the whole machinery in a given mining company. Examples of the IT tools aiding in the management of machines, used in selected Polish mining companies are presented further in this paper.

2.1. *Machinery management systems in hard coal mining*

A system called SZYK-MASZYNY is used in all the mines of the Katowice Holding Company and Jastrzębie Coal Company to manage mining machines and equipment, their components and spare parts. The system includes a central library of machines, based on the ORACLE database. The database contains information on the particular machines and their components, i.e. specifications and technical drawings, documentation, photographs and various user-defined parameters relating to the individual machines [1].

The system works with the RFID (Radio Frequency Identification) reader using a low-power radio signal to transmit data between a tag and the reader (Figure 1). An integral part of the system is the GATHER software developed by the Centre for the Mechanization of Mining (KOMAG). The software runs in the client/server architecture and consists of the MS SQL Server 2000 and the GATHER client application. A central system database, whose structure is being developed by liner manufacturers and users and by repair firms, is under construction.

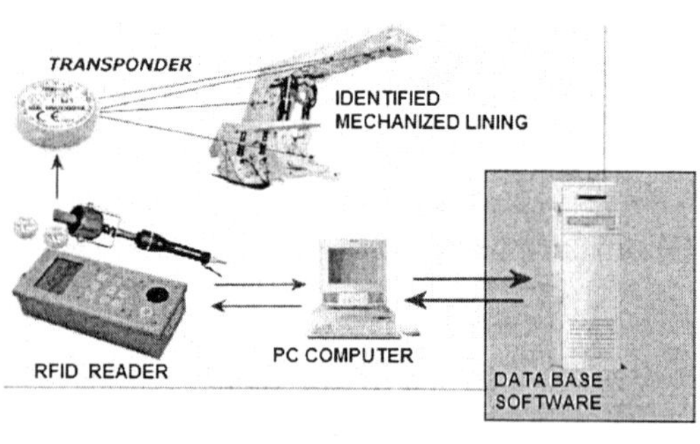

Figure 1. Mining machine components identification system based on RFID technology [3]

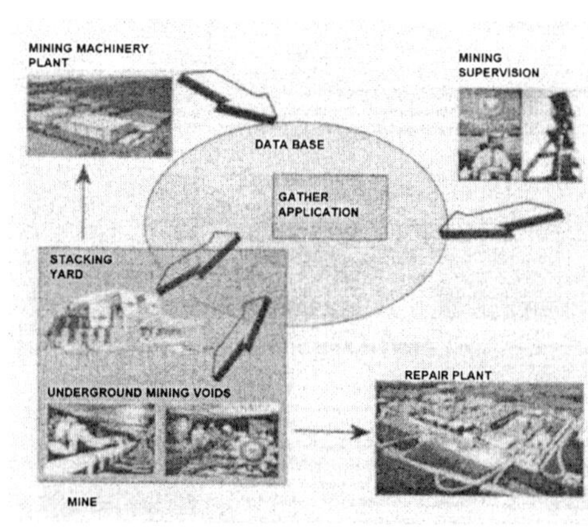

Figure 2. Information structure of mining machine and equipment components identification system [4]

In order to determine the degree of wear of machine components special algorithms were introduced into the system. Also mining supervision (at different levels) plays a decision-making role in the system. The information structure of the SZYK-MASZYNY system is shown in Figure 2. The system was developed jointly by the Z.U.E. ELSTA Company, the Mining-Metallurgical Academy (AGH), KOMAG and the Silesian Polytechnic. The system helped to solve the problem of objective and reliable identification of machine and equipment components and recording of their working life and operating conditions under the materials management in the mines and it made it possible to process the data in order to assess the degree of wear of the particular machine components [2], [3], [4].

2.2. *Machinery Management Systems in KGHM "Polska Miedź" S.A.*

A system of mining machinery management, called EKSPERT^{SMG}, which supplies information on the current costs, the performance of the production tasks and assessments of the efficiency and effectiveness of managing the self-propelled mining machines and their subassemblies, works in KGHM "Polska Miedź" S.A. (the Mining & Metallurgical Copper Conglomerate Polish Copper PLC). The main goal of the system is to standardize the cost accounting procedures for each task carried out in the Conglomerate and configure the SAP/R3 system. Currently, the system works in data warehouse technology enabling the optimization of the generation and distribution of information within the three mines (where over 1280 various mining machines work) and in the KGHM "Polska Miedź" S.A. Management Office. The use of data warehouse technology ensured quick access to data, flexibility in the development of data grouping and conversion rules, and an effective user-friendly graphic interface in the form of a web browser. The EKSPERT^{SMG} system enables, among other things, the evaluation of the work and productivity of the operators of the self-propelled mining machines, the display of machinery utilization, the evaluation of the productivity of the self-propelled mining machines by the departments ordering the machine services, the analysis of the services performed by outside firms, the presentation of direct and indirect costs relating directly to the self-propelled mining machines and the display of production realization [2], [5], [6]. Figure 3 shows the main components of the EKSPERT^{SMG} system.

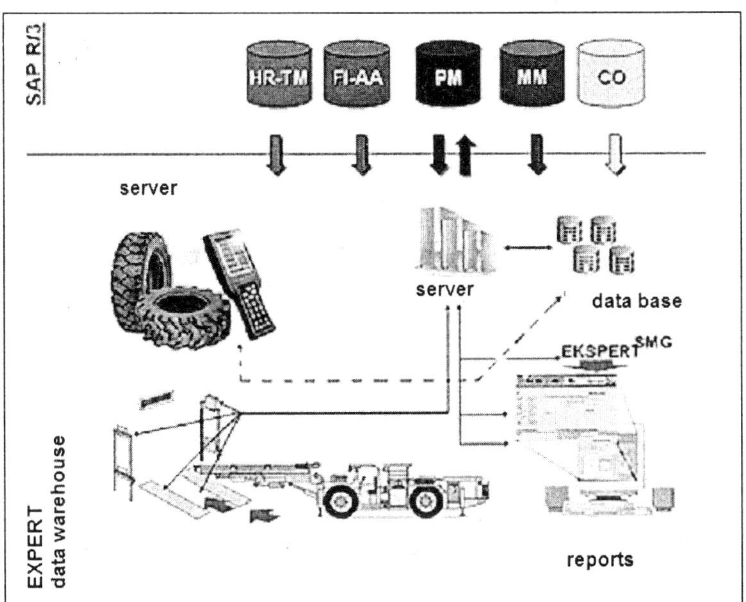

Figure 3. Main components of mining machinery management system EKSPERT^{SMG} [2]

3. BELT CONVEYOR MANAGEMENT

In the previous chapter a survey of mining machinery maintenance management solutions was made. In the authors' opinion (based on an analysis made for the lignite opencast mines), it is necessary to build a belt conveyor oriented computer based maintenance system. Similar ideas can be found in [7]. A more comprehensive survey of the literature in this field is presented in [8]. On the basis of design, operation, condition monitoring and other experience we propose a holistic, condition based maintenance approach which goes much beyond cost or reliability analyses. However, such a task is difficult and requires wide knowledge relating to drive units, belts, idlers, etc. The present work is a summary of many years of experience in the field of condition monitoring.

3.1. *Aims and tasks of computer based maintenance system*

Let us remind ourselves that the belt conveyor maintenance management system is a decision system aided by an information system, usually in the form of a computer decision support system (DSS) [9].

Besides inventorying objects, documentation and technical data, keeping records of and analyzing work effectiveness/time (output, run time, standstills), recording and analyzing replacements, malfunctions, failures and periodic and preventive maintenance, analyzing and planning (major) materials and spare parts management, keeping records of personal data (including monitoring the qualifications and professional licences of the machine operators), the primary belt conveyor maintenance management tasks (acc. to [9]) include: preparing scenarios, strategies or individual decisions concerning maintenance.

3.2. *Structure of belt conveyor*

The belt conveyor consists of a drive unit (a motor, a clutch and a transmission gear), a system of (head, tightening and tail) pulleys, a belt loop, idlers and several auxiliary devices, such as control and support (e.g. cleaning) equipment, monitoring and diagnostic elements, etc. Figure 4 shows two schematics of the belt conveyor structure.

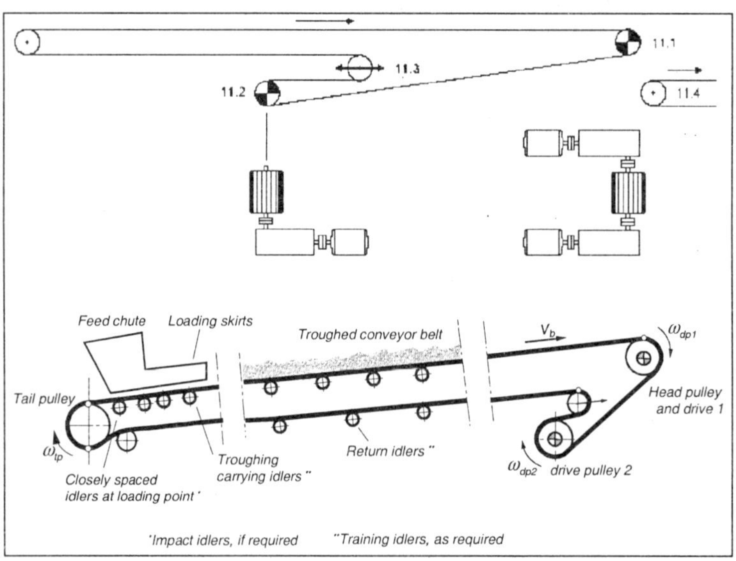

Figure 4. Structure of belt conveyor [5]

The failure frequency analysis carried out for the particular system components in [10] indicates that the key belt conveyor components are: the belt, the idlers, the pulleys and the drive units. Thus data on the operation of the above components are essential for belt conveyor maintenance management.

3.3. *Source of Data for management system*

The key question during management systems design concerns the kind, format and source of acquired data. Figure 5 shows a diagram, according to subjects and sources, of information needed to build a belt conveyor maintenance management system.

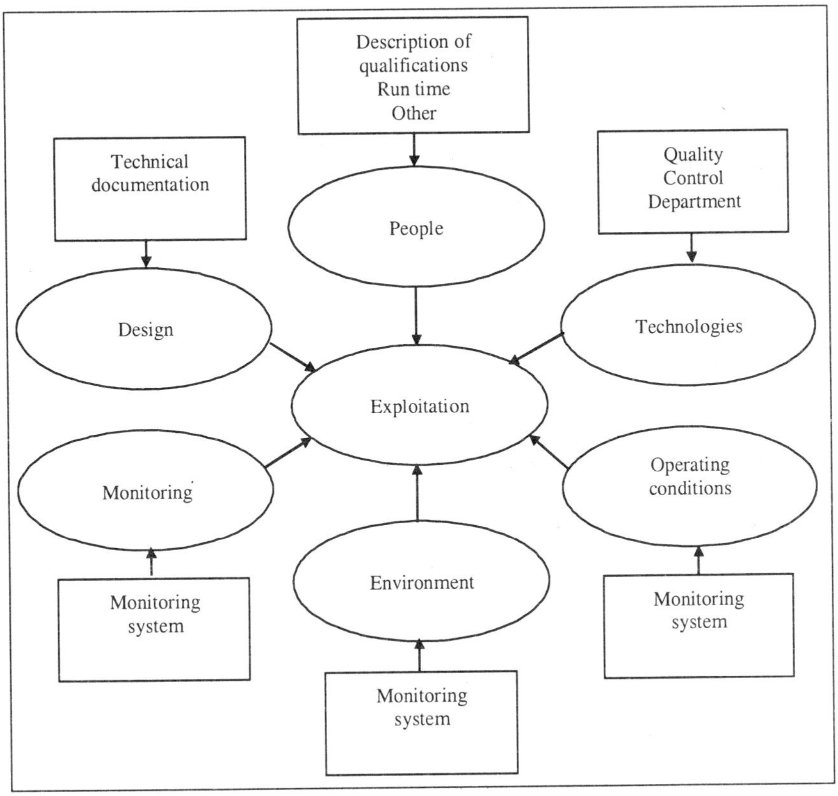

Figure 5. Data flow diagram, taking into account data types and sources

Since the following factors:

- the adopted design solutions,
- the machines manufacturing technologies and the mining technologies,
- the mine environment,
- the human factors,
- degradation processes resulting in changes in the operating parameters of the machines in comparison with the design assumptions,
- the process parameters,

have a significant effect on the operation of mine machinery, they must be taken into account in the belt conveyor maintenance management system.

It seems that the graphic and descriptive database should have a structure in which tables would apply to the key belt conveyor components (i.e. the motor, the transmission gear, the pulleys, the belt, the rollers and other components, Fig. 6) and relationally linked to other tables according to the concept shown in Figure 5.

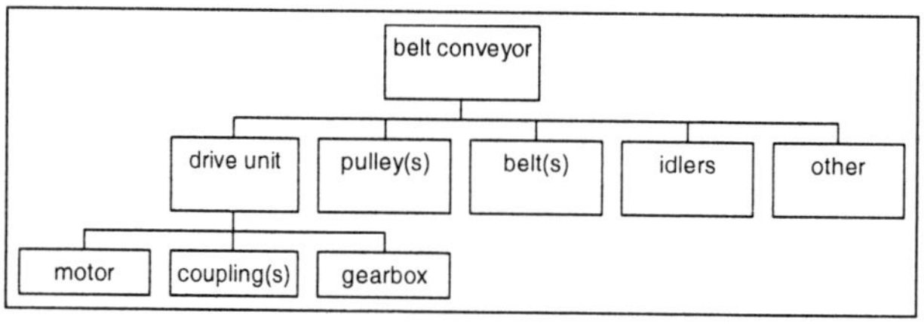

Figure 6. Object-oriented structure of database

An example of the detailed structure of information relating to the pulley is shown in Figure 7.

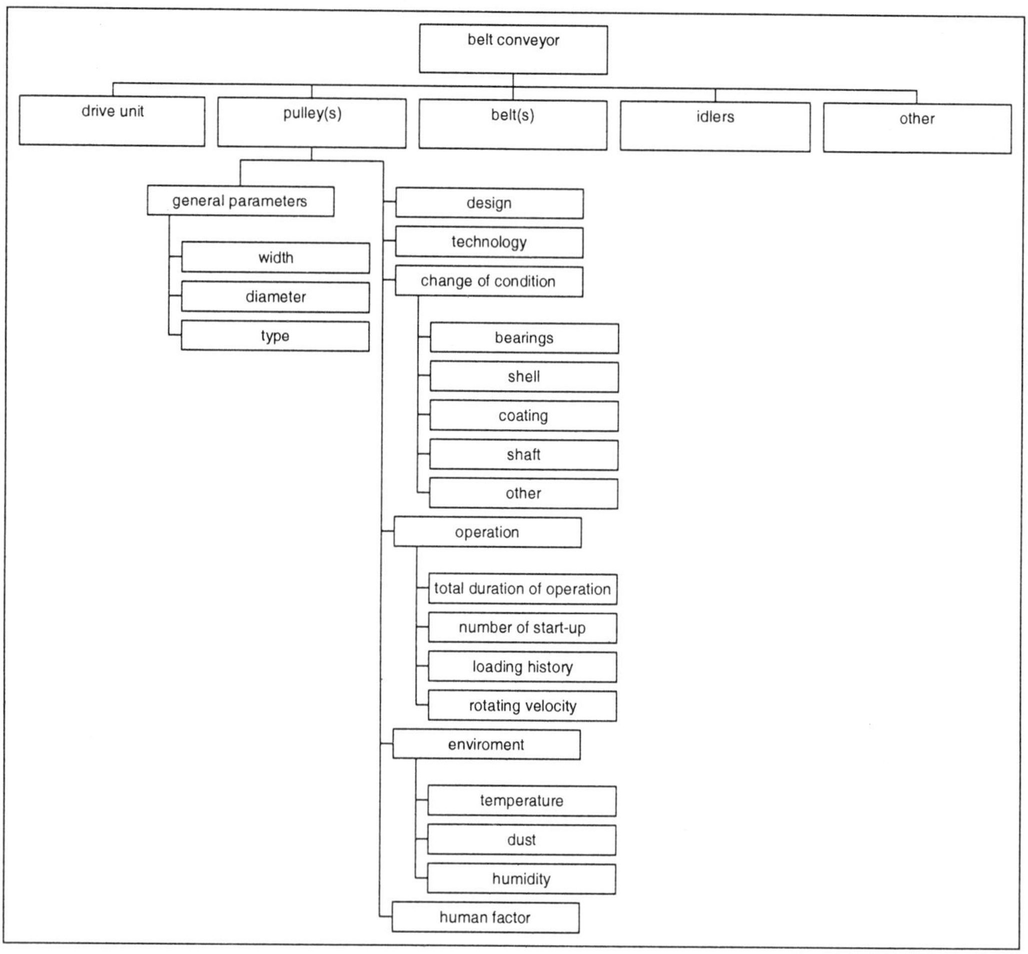

Figure 7. Detailed information relating to pulley

3.4. *GIS systems*

The continuous increase in the amount of data, the importance of information possession and the great popularity of computer databases stimulate interest in techniques of automatic, intelligent and fast data processing aimed at acquiring knowledge about objects described by data contained in a database. Geographic information systems (GIS), designed for acquiring, verifying, storing, integrating, 3-D analysis and visualizing data spatially related to the Earth's surface, in which each element is described by coordinates X, Y or (optionally) Z, occupy a special place in the class of database systems. The structure of GISs geared towards spatial analyses is shown in Figure 8. Such analyses are made through system-predefined reports, statements and charts with 3-D visualization, or using the structured query language (SQL) to query integrated graphic and descriptive databases. The graphic database is made up of numerical thematic maps, orthophotomaps and numerical terrain models. The particular graphic database objects are linked to the descriptive database whose contents depend on the kind and range of information stored in the database. Currently such systems are used in areas where the considered problems are particularly complex and in order to solve them large data sets coming from different sources need to be acquired and processed.

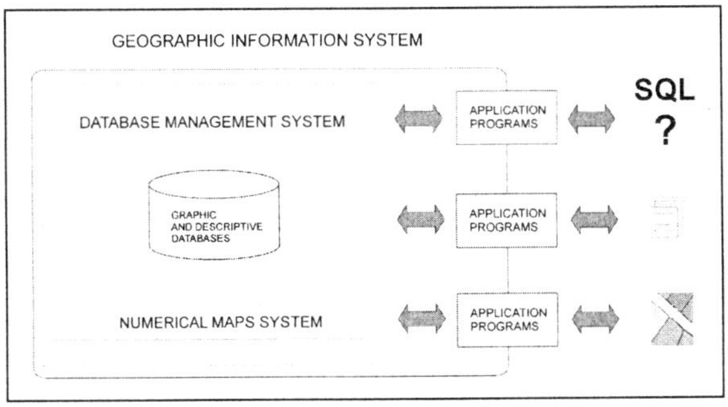

Figure 8. Structure of GIS

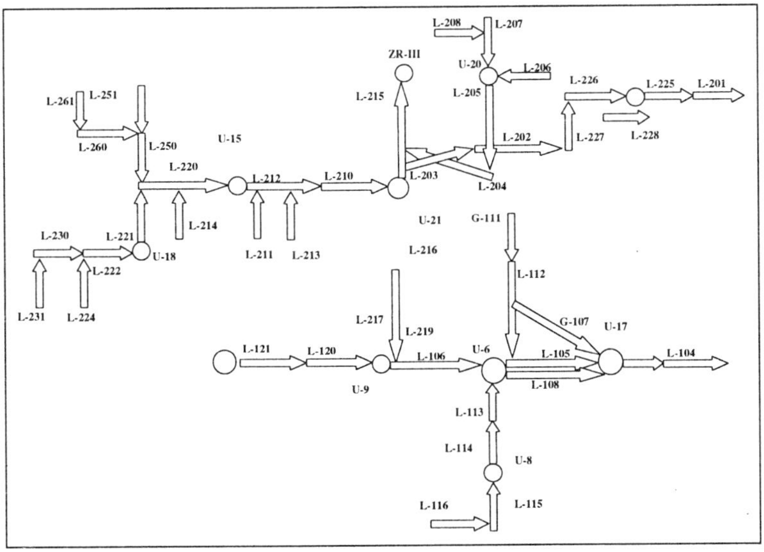

Figure 9. Scheme of belt conveyor transport system in underground mine

Figure 9 shows a scheme of the conveyer transport system in an underground mine. Each arrow represents a belt conveyor with a drive unit, pulleys, belts, idlers, etc. (Fig. 10).

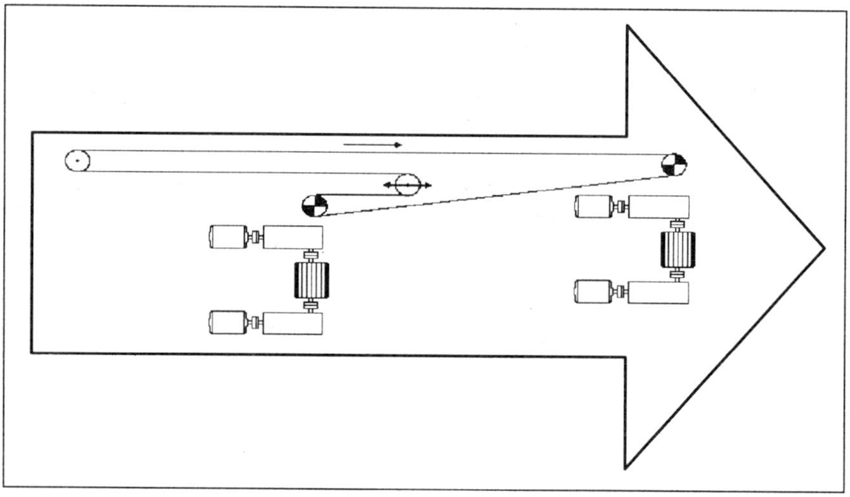

Figure 10. Detailed content of each arrow

When one of the arrows is double clicked, a simple table with basic information relating to the selected conveyor appears. Data exploration is discussed in the next section.

3.5. *Data analysis techniques*

The huge number of data requires advanced automatic procedures for the different data analysis techniques, such as filtering, sorting, classification, etc. The application of the statistical data analysis approach [11] or the more advanced, data mining approach [12] with neural networks [13], fuzzy systems and so on will start a new chapter in mining machinery management. By combining different data relating to the factors discussed above one can build procedures for detecting and diagnosing faults and forecasting the fault-dependent lifetime for the particular objects. This will constitute a basis for maintenance decisions. Understanding the origin of a fault – the so-called root cause analysis introduced by Bartelmus [14] – is of vital importance. Szymanski [15] identified the main problems connected with the operation of the drive units used in conveyor systems. He showed that depending on the type of coupling, the percentage and type of faults differ greatly.

There are many interdependencies between design factors and machine operation. For more details the reader is referred to the previous papers done by Bartelmus.

The dependence between operation effectiveness and a change in condition is obvious and does not require explanation.

It may be interesting to the readers how the environment can influence operation. In opencast mines this influence is quite clear (cold winter and hot summer, rain, etc.). It is typical for the mining industry that in some parts of the mine dust may adversely affect the machines, especially their bearings. The degradation of bearings is very high in comparison with other industries (power plants, the petrochemical industry, etc.). The main problem is excessive backlash and bearing seizure instead of inner/outer race damage.

Therefore maintenance should be treated in a holistic way. This means that one must investigate how the different factors influence operation. For this ambitious task advanced data processing and analysis techniques and deep understanding of conveyor behaviour are necessary.

However, let us start with simple things. Using the GIS environment one can examine data relating to a selected conveyor. Conveyor selection is made by a simple mouse click on the object and automatically the associated database is displayed (Fig. 11).

A simple analysis can be performed through database queries using the SQL wizard. For example, the query (the criterion for data selection/filtering) can be "show belt conveyors with breakdown/repair in the last 5 months". As a result, one will see conveyors marked with light blue dots, (Fig. 12).

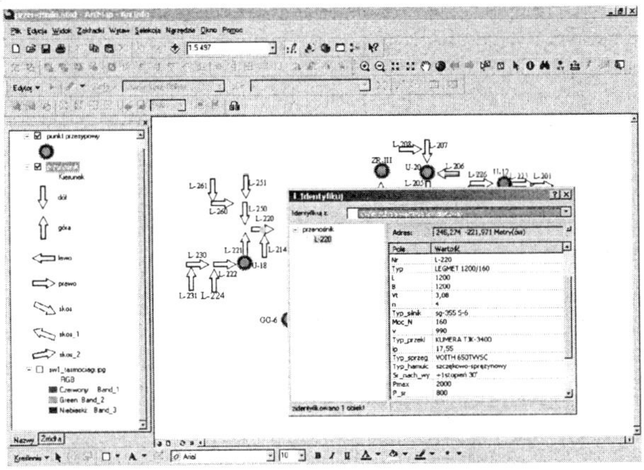

Figure 11. Displaying information relating to selected conveyor in GIS environment

For the selected data a report, in the form of a table or chart (Fig. 13), can be produced. From the maintenance point of view, the type of event is important. The first kind of event is a replacement done during a scheduled stoppage or when a repair had been planned and the problem was fixed during a scheduled stoppage. The second kind of event is an unexpected breakdown as a result of which production is lost. The above types of events are typical for the mining industry where stoppages due to, for example, rock blasting, the lack of the CM approach, the use of machines with shorter lifetime frequently occur (degradation processes are quick and different than in other industries).

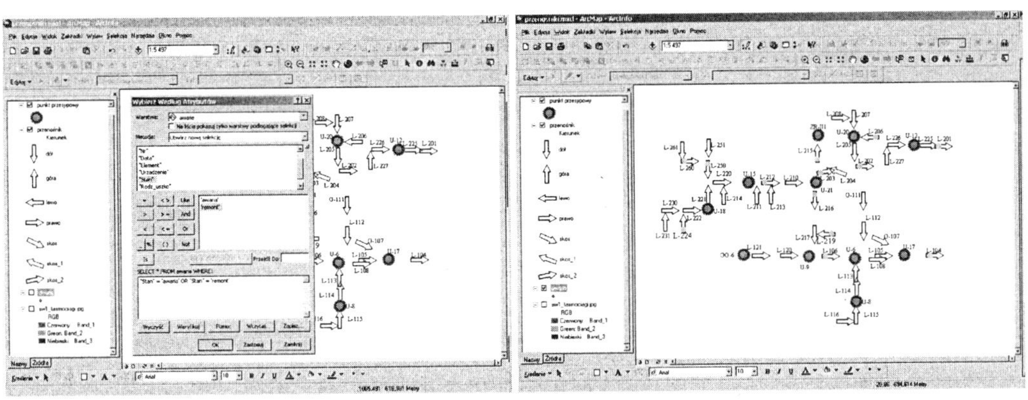

Figure 12. Example of data base query (SQL wizard) and filtering results

The most dangerous and expensive are unpredicted breakdowns (the maintenance staff is usually not ready for repairs, etc.). One may notice that breakdowns are the most frequent type of events for every type of elements. This means that the current maintenance strategy is probably inadequate.

Replacements during the scheduled stoppages are not frequent. This indicates that it is difficult to recognize a fault by visual inspection without using any CM methods. The third type of event: "minor repairs" is a less critical form of breakdown – the transport system is stopped for repair at a convenient time to minimize the impact on the production output.

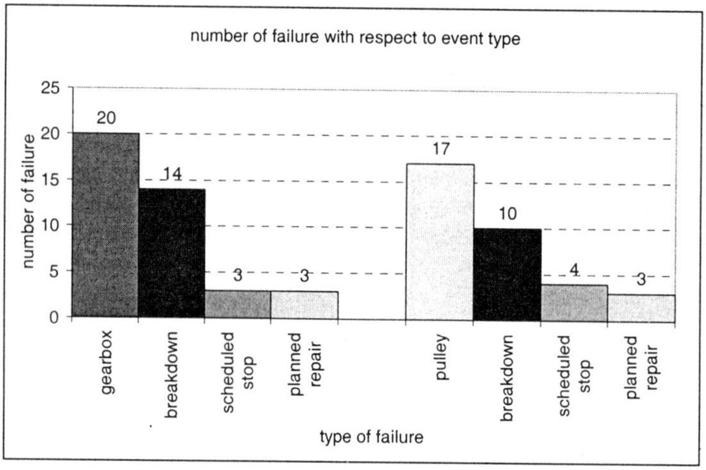

Figure 13. Exemplary results of analysis

CONCLUSION

The effective operation of a conveyor system requires a suitable maintenance policy. Unfortunately, in many cases in the mining industry (especially in underground mines) condition based maintenance policy is still not well developed. In order to improve this situation, this preliminary research was carried out. The use of GIS's to aid in belt conveyor maintenance management was discussed. The effectiveness can be improved by implementing machinery management procedures. The lack of tools for belt conveyor management was pointed out. Attention was drawn to the peculiar nature of the mining sector, which makes it difficult to directly apply the solutions existing in other sectors. The GIS platform was proposed as the environment for building belt conveyor maintenance management tools. A concept of building such a system was formulated and the latter's principal assumptions and data sources were defined. A new philosophy based on the investigation of the influence of the design, the technology, the operating conditions, the change in condition, the human factors and the environmental impact was proposed. Simple examples showed that it is necessary to think about maintenance in a holistic way, which entails the investigation of how the different factors influence operation.

REFERENCES

[1] Rogala J., Jenczmyk D., Bryk M., Helbin K. 2007: Współpraca systemu identyfikacji elementów sekcji obudowy zmechanizowanej z oprogramowaniem SZYK-MASZYNY. Materiały Szkoły Eksploatacji Podziemnej 2007, str. 849–857.

[2] Kicki J., Dyczko A. 2006: Szanse technologii RFID w polskim przemyśle wydobywczym. Nowe Górnictwo. nr 1/2006, www.nowegornictwo.pl

[3] Fitkowski K., Jankowski H., Krzak Ł., Stankiewicz J., Szczurkowski M., Warzecha M., Worek C. 2005: System identyfikacji elementów maszyn górniczych z wykorzystaniem technologii RFID. Materiały z Poznańskich Warsztatów Telekomunikacyjnych, 8–9 grudnia 2005.

[4] Szczurkowski M., Jankowski H., Worek C., Maksymowicz L., Meder A. 2008: Praktyczne doświadczenia wdrożenia w zakresie wykorzystania technologii RFID oraz nowoczesnych systemów bazodanowych do ewidencji części maszyn górniczych. Materiały Szkoły Eksploatacji Podziemnej 2008, str. 363–372.

[5] Kicki J., Dyczko A., Śliwiński P. 2006: Stan aktualny i strategia rozwoju systemu gospodarki maszynami górniczymi EKSPERTSMG. Materiały Konferencyjne Szkoły Eksploatacji Podziemnej 2006.

[6] Materiały KGHM "Polska Miedź" S.A.

[7] Wędrychowicz D., Kulinowski P.: A Concept of a Belt Conveyor Maintenance Database. Transport Przemysłowy i Maszyny Robocze, No. 2/2008.

[8] Gorniak-Zimroz J., Zimroz R., Król R., Jurdziak L.: The Application of GISs to Support Belt Conveyor Maintenance Management. Prepared for the Third International Congress on Design and Modelling of Mechanical Systems, Hammamet – Tunisie, Tunisia 2009.

[9] Dałkowski B.T., Jurdziak L., Kawalec W.: Personal Computer System for Maintenance Management of Belt Conveyor Systems. Proceedings of the XXII International Symposium on the Application of Computers and Operations Research in the Mineral Industry (APCOM'90), 1990.

[10] Zimroz R., Król R.: Failure Analysis of Belt Conveyor Systems for Condition Monitoring Purposes. Prepared for the Third International Congress on Design and Modelling of Mechanical Systems, Hammamet – Tunisie, Tunisia 2009.

[11] Jurdziak L., Stolarczyk R., Zawadzka B.: New Tools for the Computer Aiding of Belt Maintenance Management. Transport Przemysłowy, Vol. 2(8)/2002.

[12] Górniak-Zimroz J., Woźniak J., Zimroz R.: The Potential for Data Mining in Geographic Information Systems Geared towards Earth Resources Management. Górnictwo i Geologia VIII, Oficyna Wydaw. PWroc., pp. 75–86, Wrocław 2005.

[13] Bartelmus W., Zimroz R., Batra H.: Gearbox Vibration Preprocessing and Input Values Choice for Neural Network Training. AIMETH – Artificial Intelligence Methods 2003.

[14] Bartelmus W.: Computer-Aided Multistage Gearbox Diagnostic Interference by Computer Simulation. University of Tech. Press, Wrocław 2002.

[15] Szymanski Z.: Reliability Investigation of Mines Machine Drive Motors. Zeszyty Problemowe – Maszyny Elektryczne, Nr 78/2007, Politechnika Śląska, Gliwice 2007.

International Mining Forum 2009, Sobczyk, Kicki & Saługa (eds) © 2009 Taylor & Francis Group, London, UK, ISBN 978-0-415-80428-8

Method of Copper Deposit Mineralization Discrete Modelling With Use of Crystal Ball Application Based On Copper and Silver Grades Examples

Michał Kopacz
MEERI of the Polish Academy of Sciences, Krakow, Poland

ABSTRACT: The fundamental idea of simulation is replacing unknown variables with their statistical distributions. Selecting distributions fitting geological data collected for example describing variation of geological parameters such as copper and silver grades, reveals to be difficult. There are examples of stochastic, random modelling both with use of continuous or discrete distributions. The reason for inexistence of a simple solution is usually unpredictable; in this case it is a complicated character of the phenomenon that cannot be described with known mathematical function.

The paper shows the method developed by the author allowing for distribution fitting to unknown (decision) variables which are basically discrete. Researches have been done on examples of statistical modelling of copper and silver grade in "X" copper ore deposit. Discrete methods are suggested to be used in case of modelling smaller collections of observations, which is very common at an initial stage of ore development. Being conscious of input data limitation, the attempt to modelling with the best fitted discrete distribution was supported by Crystal Ball and Statistica software. The source of geological assays was exploratory boreholes (1,5×1,5 km blocks). As a result of estimating, probability distributions of metals grade (Cu, Ag) were obtained as a weighted sum of uniform distributions.

The evaluated method seems to be simple and reliable. The precision achieved depends on density and quality of geological data. The method allows also for avoiding the problem of limited data while maintaining an acceptable level of agreement with the actual. Along with growth of data availability the precision of the method remains high and resultant discrete distributions resemble continuous ones. The method can be also applied to other scientific areas. The only condition is to have defined sets of data and software allowing for relative probability estimating (for instance – Crystal Ball).

KEYWORDS: Metals (Cu, Ag) grade, simulation models, method of distribution fitting, statistical data processing, Crystal Ball software

INTRODUCTION

Difficulty or, in some cases, impossibility to use classical continuous or discrete modelling enhances the need for utilization of alternative methods. Simulation models paradigm, in which all or sometimes only a few of unknown (decision) parameters are replaced with their distributions, has been undertaken in publications of Cyrnek, Magda, Snopkowski, Jasiewicz, Głodzik, Woźny, Jurdzik, Fisher, Karbownik, Jabłońska-Firek, Saługa, Tumidajski and Brożek [1]. In their publications the authors use logarithmic and lognormal distribution to describe copper ore mineralization. These are therefore examples of continuous distributions. Following continuous distribution definition (even in isolation from sophisticated mathematic and probability) their implementation is restricted. To be able to use continuous forms, basic condition is to have very numerous collection of observations and also to allow for all values in a range. In our example there is only a limited number of data, so more appropriate seems to be modelling with use of a discrete (uniform) distribution. Of course, there are known methods of estimating missing data making use of a statistical method still possible [3]. Data has been verified for its representativeness, explicitness and forecasting [4]. Researches done over data input confirmed ability to apply in specific, modified way what is the core of the presented method. The topic seems to be still evitable and the purpose of the publication is to present the method and its application on practical examples.

1. METHOD DESCRIPTION – METHODOLOGY

The method presented in the paper uses statistical tools as well as simulation software (Crystal Ball) to build and value different scenarios of mining projects. In general, it relies on replacing continuous distributions with finite numbers of uniform distributions. Share of particular one is calculated based on observed probability allowing for sampling in accordance with reality in simulation models. The key is to drive the relative probability function form observed data. The main steps when using the method of evaluation are:

- Dividing the ore body into Boldyriew blocks and estimating main geological parameters (copper and silver grades, tonnage of copper and copper ore) for each of them.
- Grouping observations into classes of estimated distributions. Initially continuous distribution is used as a "pattern". In a practical calculation the number of classes is automatically established by Statistica according to the existing rules [3]. Every class if represented by its average value plus-minus 5% of allowed deviation.
- Grouping observations means classifying each Boldyreiw block to one distribution's class and calculating relative probability function. Block sum of copper ore divided by total ore mass defines probability function. Total sum of probability equals 1.
- In last step calculation output is taken into Crystal Ball as input to distribution fitting menu. Lack of data is being considered as 0 and exemption of sampling.

Table 1. "X" ore body presentation – geological data

	Parameter	Measure	Boldyriew Block characteristic			
			Lithographic series			Total
			Carbon	Shale	Sandstone	
1.	Ore body surface	m^2	39,214,189			
2.	Average thickness	m	0.55	0.35	1.45	2.06
3.	Copper ore resources	tonne	57,047,160	30,964,553	118,593,741	206,605,454
4.	Copper grade – average	%	1.36%	6.85%	1.33%	2.17
5.	Silver grade – average	g/tonne	50.4	246.6	47.7	78.42
6.	Copper grade equivalent	%	1.86%	9.32%	1.81%	2.95
10.	Total copper	tonne	775,271	2,120,762	1,579,669	4,475,702
11.	Total silver	kg	2,875,177	7,635,859	5,656,921	16,167,957
12.	Total copper equivalent	tonne	1,062,789	2,884,348	2,145,361	6,092,497

2. "X" ORE BODY PRESENTATION

"X" ore body had been recognised from 27 drilled boreholes inside and close to ore boundaries. Given copper deposit is recognised as discontinuous and irregular occurring among various seam thickness. Variation over mineralisation and composed geological structure has been confirmed. In most wells lithographic series is composed of 3 different rock types: carbonate, copper shale and sandstone. Average seam thickness of each Boldyriew block (Tables 1 and 2) varies between 1.4––3.14 m. Arithmetic average copper grade equals 2.17%, and silver – 78.4 g/tonne.

Within the range of a mining face (ore thickness) dilution was established upon real waste rock estimates. Over the first step extraction face total dilution estimation was fixed at 6%. Total copper and silver amounts were calculated based on actual measurements. Extended ore specification was presented in Table 2.

Table 2. "X" ore body presentation – extended geological data

Block	Thickness [m]	Dilution [%]	Cu grade [%]	Cu grade – dilution corrected [%]	Total Cu [1000 tonnes]	Ag grade [g/tonne]	Ag grade – dilution corrected [g/tonne]	Total Ag [tonnes]	Cu grade – equivalent [%]	Total Cu corrected [%]
X1-1	1.79	6	4.28	4.03	441.99	150.30	141.28	1,550.68	6.17	6.06
X1-2	1.56	21	4.42	4.07	486.09	139.50	128.34	1,534.49	6.17	6.03
X1-3	1.78	9	3.90	3.54	164.32	131.40	119.43	554.04	5.54	5.39
X2-1	1.98	6	2.89	2.72	108.43	111.60	104.90	418.87	4.29	4.20
X2-2	1.48	25	3.54	2.63	396.95	128.70	95.52	1,444.36	5.15	4.73
X2-3	2.02	6	2.81	2.64	109.78	85.50	80.37	334.25	3.88	3.82
X3-1	1.65	16	2.29	1.91	76.16	58.50	48.95	194.91	3.02	2.90
X3-2	1.70	14	2.14	1.85	83.50	77.40	66.91	301.73	3.11	2.98
X3-3	1.40	30	1.93	1.34	220.90	79.20	55.26	908.38	2.92	2.62
X3-4	2.22	6	2.53	2.38	185.23	83.70	78.68	613.04	3.58	3.52
X3-5	2.19	6	2.32	2.18	226.37	88.20	82.91	859.87	3.43	3.36
X3-6	1.97	6	1.83	1.72	98.21	69.30	65.14	372.52	2.70	2.64
X4-1	2.43	6	1.71	1.61	72.52	68.40	64.30	290.07	2.57	2.52
X4-2	2.78	6	1.78	1.68	73.44	46.80	43.99	192.89	2.39	2.35
X4-3	3.16	10	1.68	1.51	173.41	49.50	44.55	510.03	2.30	2.24
X4-4	2.95	8	1.57	1.44	197.23	55.80	51.34	702.78	2.27	2.21
X4-5	2.34	6	1.79	1.68	188.17	54.00	50.76	567.34	2.47	2.43
X4-6	1.65	16	1.08	0.90	60.23	53.10	44.43	296.14	1.75	1.64
X5-1	1.89	6	0.88	0.83	38.37	58.50	54.99	254.46	1.62	1.57
X5-2	2.12	6	0.78	0.74	64.31	48.60	45.68	399.19	1.39	1.36
X5-3	2.45	6	0.90	0.85	29.87	52.20	49.07	173.27	1.55	1.52
X5-4	1.98	6	0.81	0.76	53.02	42.30	39.76	276.86	1.34	1.31
X5-5	1.67	15	0.69	0.53	68.31	42.30	35.86	458.62	1.16	1.08
Average, Sum	*2.06*		*2.17%*		*3,717.95*	*78.42*		*13,447.42*		

3. APPLYING METHOD – EXAMPLES

3.1. *Estimation of copper grade distribution*

Practical aspect of copper ore excavation supports the rule that copper grade is usually higher during extraction phase than development. The reason is – higher dilution amount caused by exploitation apart from located ore body. The increase of dilution means higher metal contamination and very often – less of final copper and silver production. This observation was reflected in calculation of inputs by 25% reduction of copper ore amount in every distribution class (distribution estimated for development phase). Although less metal output implicates lower production economic effect it doesn't cause shifts in relative probability function. So, the estimation is still accurate.

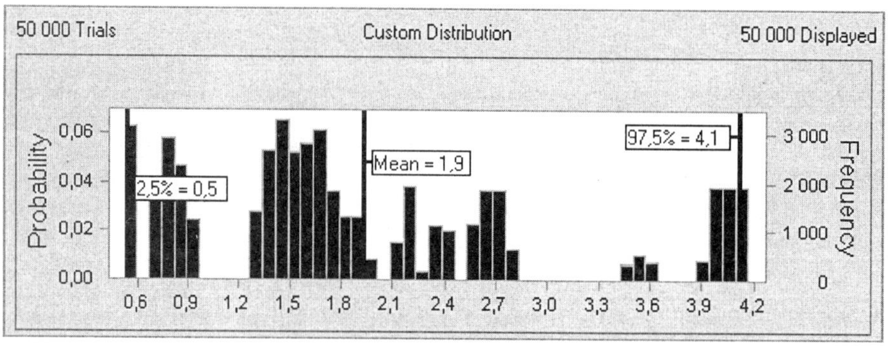

Figure 1. Copper grade distribution – extraction phase [%]. Source: own study

Table 3. Data input used for copper distribution fitting – exploitation phase

Distribution class	Minimum in class grade [%]	Maximum in class grade [%]	Relative probability function [%]	Observations in classes	Total copper ore amount [1000 tonnes]
1	0.52	0.55	6	1	10,842
2	0.72	0.78	9	2	14,759
3	0.81	0.87	4	2	7,669
4	0.88	0.93	3	1	5,577
5	1.31	1.38	7	1	11,470
6	1.40	1.47	7	1	12,595
7	1.48	1.55	6	1	10,304
8	1.57	1.74	14	4	24,244
9	1.75	1.96	7	3	12,531
10	2.13	2.24	6	1	9,749
11	2.32	2.44	4	1	7,324
12	2.56	2.78	11	3	18,885
13	3.45	3.63	2	1	4,216
14	3.93	4.17	12	2	21,317
Total			100	24	171,483

To replicate natural copper ore body mineralisation, in advance, discrete distribution was established. This distribution can be also considered as continuous in some of date range. Theoretical aspect and steps of distribution fitting were discussed in earlier section. Relative probability function

(Tables 3, 4, 5 and 6) was calculated by dividing copper total mass in every distribution class and total amount of copper ore. Minimum and maximum of every distribution class were established as average in class +/–5% of allowed deviation.

Finally, two different distributions types were constructed both for development and extraction phases. Figure 1 presents histogram of copper grade variation for exploitation phase. Weighted average metal content equals 1.93%. Table 3 shows data input used for fitting.

Figure 2 presents second of mentioned distribution. "White gaps" reflect classes with probability function "0", excluded from sampling.

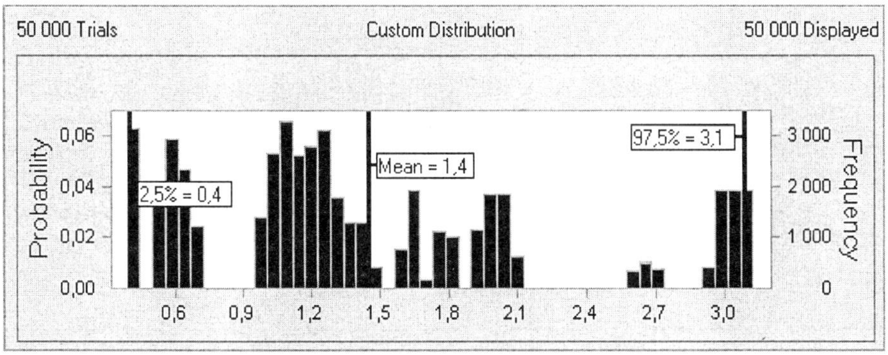

Figure 2. Copper grade distribution – development phase [%]. Source: own study

Table 4 includes data input for distribution presented on the Figure 2. The same calculation rules were conducted to achieve shown histogram. Column 4 reflects relative distribution function values [%]. For example; row 8. column 4 – 14% means – statistical chance that every of 14 from 100 sample would be taken from the range 1.18–1.31% of copper grade.

Table 4. Data input used for copper distribution fitting – development phase

Distribution class	Minimum in class grade [%]	Maximum in class grade [%]	Relative probability function [%]	Observations in classes	Total copper ore amount [1000 tonnes]
1	0.39	0.41	6	1	8,132
2	0.54	0.59	9	2	11,069
3	0.61	0.65	4	2	5,752
4	0.66	0.70	3	1	4,183
5	0.98	1.04	7	1	8,602
6	1.05	1.10	7	1	9,446
7	1.11	1.16	6	1	7,728
8	1.18	1.31	14	4	18,183
9	1.31	1.47	7	3	9,398
10	1.60	1.68	6	1	7,312
11	1.74	1.83	4	1	5,493
12	1.92	2.09	11	3	14,164
13	2.59	2.72	2	1	3,162
14	2.95	3.13	12	2	15,988
Total			100	24	128,612

3.2. *Estimation of silver grade distribution*

Similarly like for Cu, silver grade histogram was estimated for conducted "X" ore body. For every distribution class relative probability function was established. Class ranges were calculated in the same way as before. Amount of copper ore and percentage Ag grade were limited by 25% in every distribution class for development phase. Silver distribution for ore extraction was shown on Picture 3 (Fig. 3), for development – in Figure 4. Detailed input parameters for statistical estimation in Crystal Ball have been presented in Tables 5 and 6.

Table 5. Data input used for silver distribution fitting – exploitation phase

Distribution class	Minimum class grade [%]	Maximum class grade [%]	Relative probability function [%]	Observations in classes	Total copper ore amount [1000 tonnes]
1	35	36.8	6	1	10,842
2	38.80	40.80	4	1	6,545
3	41.20	46.80	20	5	33,517
4	47.70	49.90	4	2	6,651
5	50.10	52.90	13	2	23,101
6	53.60	56.60	9	2	15,819
7	62.70	68.60	8	3	13,515
8	76.70	85.00	12	3	20,983
9	93.10	97.90	7	1	11,223
10	102.30	107.50	2	1	3,753
11	116.40	122.40	2	1	4,216
12	125.10	131.50	6	1	11,000
13	137.70	144.80	6	1	10,317
Total			100	24	171,483

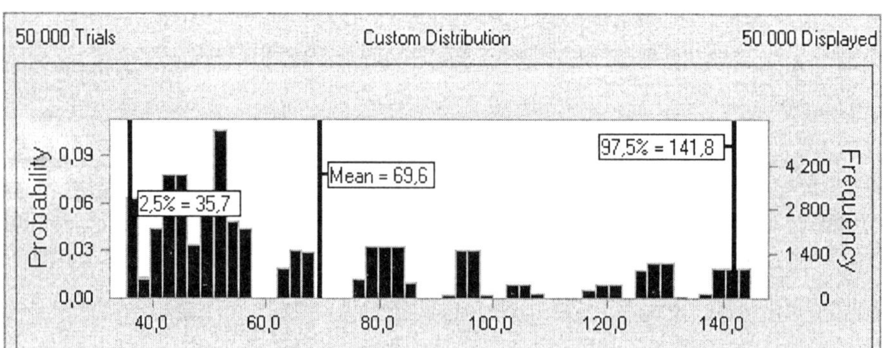

Figure 3. Silver grade distribution – extraction phase [%]. Source: own study

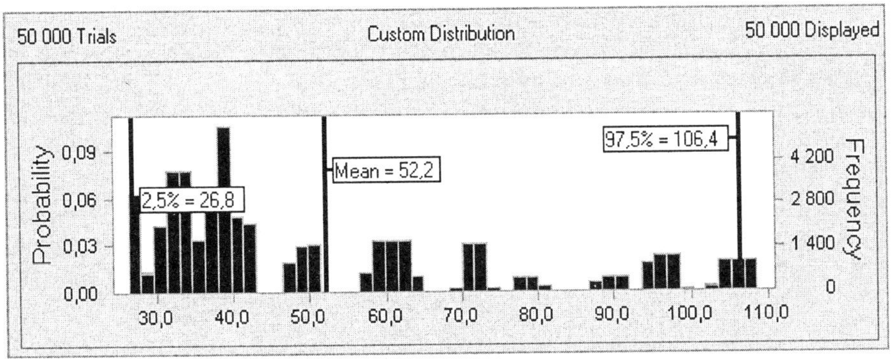

Figure 4. Silver grade distribution – development phase [%]. Source: own study

Table 6. Data input used for silver distribution fitting – development phase

Distribution class	Minimum class grade [%]	Maximum class grade [%]	Relative probability function [%]	Observations in classes	Total copper ore amount [1000 tonnes]
1	26.3	27.6	6	1	8,132
2	29.1	30.6	4	1	4,909
3	30.9	35.1	20	5	25,138
4	35.8	37.4	4	2	4,988
5	37.6	39.7	13	2	17,326
6	40.2	42.5	9	2	11,864
7	47.0	51.5	8	3	10,136
8	57.5	63.8	12	3	15,737
9	69.8	73.4	7	1	8,417
10	76.7	80.6	2	1	2,815
11	87.3	91.8	2	1	3,162
12	93.8	98.6	6	1	8,250
13	103.3	108.6	6	1	7,738
Total			100	24	128,612

3.3. *Data correlation*

Collected samples had revealed existence of positive correlation between copper and silver mineralisation in "X" ore body. The higher copper grade, the higher silver mineralisation. To allow for correlated sampling with use of Crystal Ball, fixed correlation coefficient had to be set up. Finally, correlation coefficient between copper and silver distributions was established to be 0.9. Because of character of geological researches and natural co-existence of this metals statistical relationship had to be determined. Of course, it is a kind of simplification but statistical modelling is still accurate. Random or simulation modelling with correlation implemented influences sampling rules. Main effects are listed below:

- Data dependency may cause, indirectly, dependency of third decision variables, which can be senseless and in conflict to actual relationship.
- Correlation matrix may turn to be less homogeneous. Crystal Ball software provides opportunity to correlate more than two variables simultaneously. The application informs the user of conflicting rules and suggest a matrix and coefficient modification to leave them still working. Such

"warning messages" should be treated seriously because the base of conflict lies upon invalid user assumptions. Rule of thumb is firstly – to set up the most important correlation dependences and after that others – start from the most essential.

- Correlated process influences the power of input variables on output (prediction). This hardship is caused by method used in CB to solve correlation operations. Rang correlation implemented in CB enables combining different unknown variables with others. Measuring linear dependences of variables is automated and there is no need to have the same distributions of data inputs. Utilisation of correlation coefficients driven from "raw data assays" could lead to misunderstanding and incorrect sampling. Possibility to encounter many parameters in simulation in a controlled manner is an advantage and compensation for losing some pieces of basic information.

SUMMARY

Data insufficiency is very common in practical ore body mineralization identification and establishing its value at initial stage of development. Presented examples do not deny the theory of continuous copper and silver mineralisation in experimental ore deposit but allow for sufficient and correct use of simulation techniques tailored to particular data collection. Distributions building technique presented in this paper seem to be helpful. Its powerful practical implementation and simplicity makes it very useful in practical simulation modelling. Evident advantage could be the possibility to use in case of other mathematical and statistical method (function) incompatibility. Elaborated technique can be applied in many other areas of scientific researches (not only connected with mining). And finally by method utilisation the need for statistical examination goodness of fit was excluded [8].

REFERENCES

[1] Tumidajski T.: Zastosowanie metod statystycznych w analizie procesów przeróbki surowców mineralnych. Wyd. Techniczne, Katowice 1993.

[2] Zając K.: Zarys metod statystycznych. Wyd. V, PWE, Warszawa 1994.

[3] Sadowski W.: Mała encyklopedia statystyki. PWE, Warszawa 1976.

[4] Zeliaś A., Pawełek B., Wanat S.: Prognozowanie ekonomiczne: Teoria, przykłady, zadania. PWN, Warszawa 2003.

[5] Lissowski G.: Wykłady ze statystyki: Korelacja rangowa. Uniwersytet Warszawski, 2005.

[6] Decisioneering: Crystal Ball 7.2.2 Reference Manual. Decisioneering, Denver, Colorado, 2006.

[7] Decisioneering: Crystal Ball User Manual. Decisioneering, Denver, Colorado, 2006.

[8] Kopacz M.: Metoda wyceny projektów inwestycyjnych w polskim górnictwie rudnym z wykorzystaniem symulacji stochastycznej. Rozprawa doktorska, Kraków 2007.

International Mining Forum 2009, Sobczyk, Kicki & Saługa (eds) © 2009 Taylor & Francis Group, London, UK, ISBN 978-0-415-80428-8

Valuation of Mineral Deposits in Junior Mining Companies

Krzysztof Kubacki
KGHM "Polska Miedź" S.A., Lubin, Poland

Arkadiusz Kustra
AGH – University of Science & Technology, Krakow, Poland

INTRODUCTION

The mining industry experienced the wave of consolidation in the last time. Because of lack of balance between mergers and exploration, mining companies tended to reduce exploration budget and simultaneously they started to spend much more money on mergers. It caused the trends which created a much stronger and more attractive junior companies.

The main difference remains that major diversified companies, with their focus on production, generate cash flows from metal production. Junior Companies must sell shares to get cash to maintain the exploration effort. Exploration is a systematic process that requires a lot of time, money and patience and provides high risk to investors. The life of junior mining company cannot generate new mineral properties can be very short.

In the paper it was presented the various aspects of junior mining activity with particular identification of mineral deposits treated as primary driver of value creation in junior companies. Besides it was presented estimation method of mineral deposits and some examples of mining companies with their primary mineral assets.

ASSET CONTENT REPORTING AT DIFFERENT STAGES OF EXPLORATION WORK

Companies involved in prospecting and development of mineral assets report on them according to the stage of advancing their mineral projects. This reporting could be divided into three of the following groups:

1. exploration results,
2. mineral resources,
3. mineral reserves.

All of the above reflect different levels of geoscientific confidence and different degrees of technical and economic evaluation.

Figure 1 illustrates primary definitions were taken from A guide for Reporting Exploration and Information Mineral Resources and Mineral Reserves published in 1999 by The Society of Mining, Metallurgy and Exploration Inc.

Exploration results represent early stage work undertaken on mineral projects. They are characterized by reasonably low level of geoscientific confidence. They cannot be qualified as either mineral resource or a mineral reserve. Estimates of tonnage and associated average grades are not reported along with them. However they should contain sufficient information to allow judgement of the importance of the results. It is common practice, especially for junior mining companies listed in Toronto or Sydney, to report the results of sampling or individual bore holes. These results would usually include length of intervals at which metal mineralization was intercepted at a certain cut-off grade, grades of the mineralization of both the main metal and accompanying metals, in some cases the equivalent grade of the main metal calculated by using a specified formula, and also some comments to every single bore hole reported. From the investors' point of view such results are mainly information on the work being done on a certain project, and the occurrence of certain mineralization although without the quantities and quality.

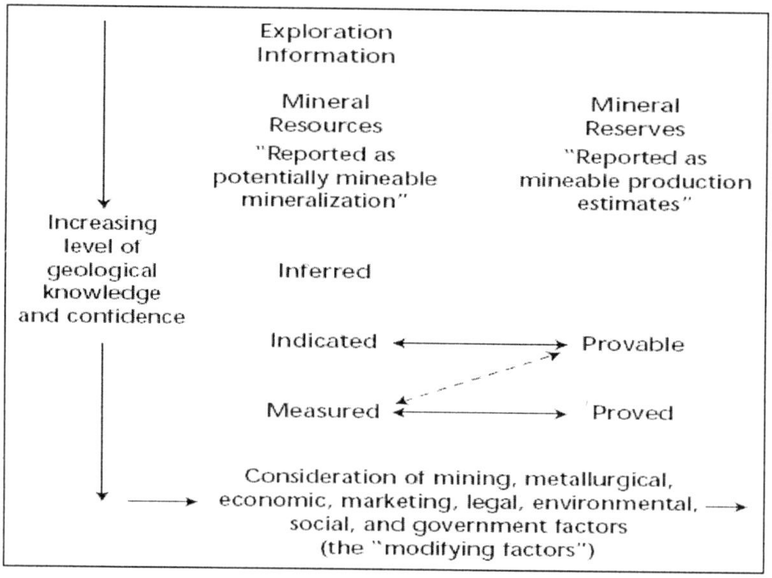

Figure 1. Risk versus increased information available on reserves/resources.
Source: A guide for Reporting Exploration and Information Mineral Resources and Mineral Reserves published in 1999 by The Society of Mining, Metallurgy and Exploration Inc.

A mineral resource is an occurrence or concentration of material of economic interest, in or on the earth's crust in such form, quality and quantity that there are reasonable and realistic prospects for eventual economic extraction.

The location, quantity, grade, continuity and other geological characteristics of a mineral resource are known, estimated from specific geological evidence and knowledge, or interpreted from a well constrained and portrayed model. Mineral resources are reported as in situ mineralization estimates.

Mineral resources are subdivided into three categories according to increasing confidence in respect of geoscientific evidence:

1. inferred,
2. indicated,
3. measured.

For inferred mineral resource tonnage, grade and mineral content can be estimated with a low level of confidence. Geological and/or grade and continuity is assumed but not verified and information gathered may be limited or of uncertain quality and reliability. This category is seldom included in economic studies (only in certain restricted circumstances in Canada).

For indicated mineral resource tonnage, densities, shape, physical characteristics, grade and mineral content can be estimated with a reasonable level of confidence. It is based on information gathered through appropriate techniques where the locations are too widely spaced to confirm geological and/or grade continuity but are spaced closely enough for continuity to be assumed. Confidence in the estimate is sufficient to allow the appropriate application of technical and economic parameters and to enable an evaluation of economic viability.

For measured mineral resource tonnage, densities, shape, physical characteristics, grade and mineral content can be estimated with a high level of confidence. It is based on information gathered through appropriate techniques where the locations are spaced closely enough to confirm geological and grade continuity. Confidence in the estimate is sufficient to allow the appropriate application of technical and economic parameters and to enable an evaluation of economic viability with a high level of confidence.

A mineral reserve is the economically mineable material derived from a measured and/or indicated mineral resource. It is inclusive of diluting materials and allows for losses that may occur when the material is mined. Dilution occurs where rock, adjacent to the ore body, is mined as part of the ore. This is common when narrow ore bodies are mined. Mining losses occur when ore is left in situ and not mined e.g. when pillars are left unmined to support the roof of underground excavations.

At this level of reporting, it is assumed that appropriate assessments, which may include feasibility studies, have been carried out including consideration of and modification by, realistically assumed mining, metallurgical, economic, marketing, legal, environmental, social and governmental factors. These assessments demonstrate at the time of reporting that extraction is reasonably justified.

Mineral reserves are divided in order of increasing confidence into:

- probable mineral reserves – for which the level of geoscientific confidence is similar to that required for the determination of indicated mineral resources;
- proved mineral reserves – for which the level of geoscientific confidence is similar to that required for the determination of measured mineral resources.

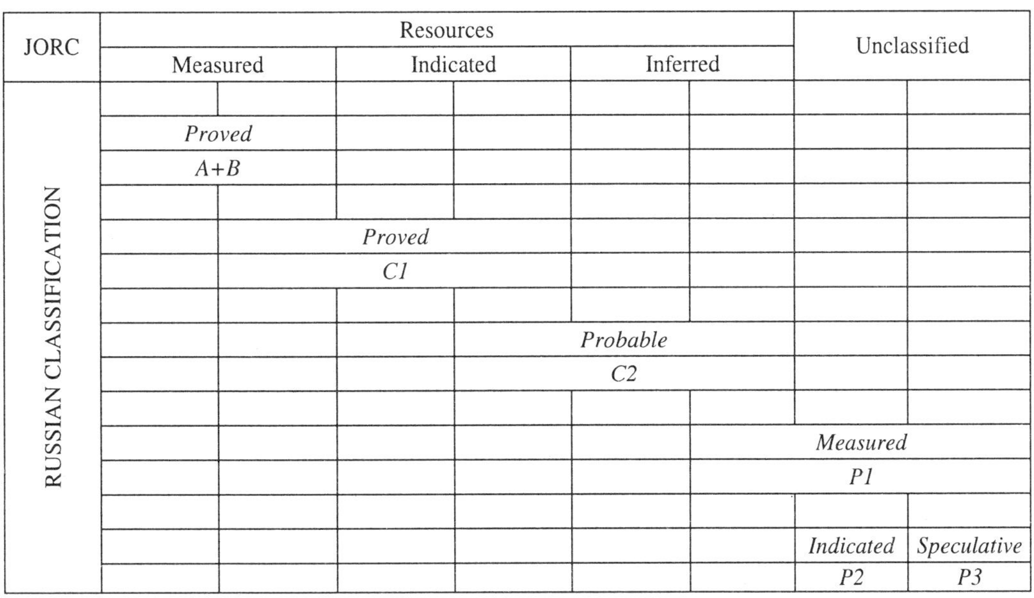

Figure 2. Approximation of Russian to JORC Classification Categories. Source: SRK Consulting

Two conventions exist where mineral resources and mineral reserves are reported for a mineral occurrence. Mineral resources can be reported inclusive of, or additional to, those resources which have been modified to produce mineral reserves. It is important to distinguish these methodologies. Many countries have also drawn up codes which set out procedures and guidelines for public reports and define the concept of the Competent or Qualified person. Thus such reports containing exploration results, mineral reserves or resources must be prepared by or under the direction of Competent (Qualified) person. This person should have a minimum of five years experience relevant to the style of mineralization and type of deposit under consideration and to the activity which that person is undertaking. The most commonly used of the regulatory codes is JORC (a code developed by the Joint Ore Reserves Committee of the Australian Institute of Mining and Metallurgy, the Australian Institute of Geoscientists and the Minerals Council of Australia), and reasonably close in most categories is the Canadian system of National Instrument 43-101. A code more complex than others and still used quite widely (in Russia and the Former Soviet Union) is the one developed in the USSR. It defines deposits under three different headings:

1. geological complexity,
2. shape of the ore body,
3. resource and reserve classifications (seven categories: A, B, C1, C2, P1, P2, and P3).

There is no approved way of equating the Russian system to JORC or NI 43-101. The Table below shows an attempt to do that, but practitioners suggest that each deposit should be completely re-evaluated under JORC rules to make them truly comparable.

SPECIFIC ACTIVITY IN JUNIOR MINING COMPANIES

The mining industry has experienced an extended period of consolidation. As companies merged, they tended to reduce the exploration budget for the new company to a level less than the combined budgets of the merged companies. This meant that experienced people were retired or got the pink slip, and many highly qualified individuals, including seasoned exploration geologists, became available as consultants to or officers of Junior Companies.

Junior companies are the combination of people, business models, property selection and their various exploration philosophies. But they involve a far more complex analysis and a measured approach to the fundamental components of the mineral exploration business, those being cash and their origin, how innovative is the geological talent, what are quality properties and creative ideas? The business model is another complex issue. It is difficult to say what factor is most important; however, nothing but dreaming results with no cash, a reality which holds true for both major and Junior Companies.

Junior mining companies operate to find exploration properties that show enough promise to attract attention of other mining companies. Ideally one of these companies will ultimately agree to joint venture or buy the properties or take over the entire company at premium, thereby increasing shareholder value and solidifying the reputations the managers' ability to generate return to shareholders. A successful takeover would enable the managers to start new company by attracting new investors who believe the same group of managers can repeat the process.

In junior mining companies with particular skills sets in finding prospective assets, possibly the highest period of risk is associated with the actual development of the asset into a producing mine. A very wide range of factors will determine success or failure at this stage, most important of which are:

- Security of tenure, given the geographic spread of junior miners in the so called "new frontier" regions with higher levels of political risk.
- Key characteristics of the orebody, particularly grade.
- Operating parameters of open pit versus underground mine development (the graphic below shows, by way of example, the implication of a steeply dipping ore body in open pit mining.
- Plant and metallurgy efficiencies (particularly in commodities that need to be separated out from by-products found in the same orebody, such as platinum group metals).

- Logistics, particularly when a mine is located in a remote "new frontier" region.
- Management record in developing projects to production stage as opposed to finding new deposits – the two stages of the lifecycle of a mine project can require dramatically different skills.
- Environmental issues that demand increasing amounts of company time and resource, covering not only the physical environment (where actual legislation will govern company requirements) but also the socio-cultural component (where local communities are capable of stalling potentially viable ventures if their concerns are not addressed).

Success and value creation by junior miners will depend overwhelmingly on:

- Large mineralisation and reserves.
- Good convertibility of resources to reserves.
- Downstream technical and economic viability.
- Management team's ability and track record in operations.
- Security of tenure, taxation and other political and fiscal issues.

VALUATION OF MINERAL ASSETS IN JUNIORS

Valuing exploration and development companies in the resource sector is fraught with challenges, even in the case of those who have access to all of the data on the asset, due to the variability of the mineral information that becomes available through the exploration and development phases, and the fact that each mineral deposit is in some ways unique in its characteristics. For investors with relatively limited access to internal company information, and usually no opportunity for a site visit, the process is even more challenging.

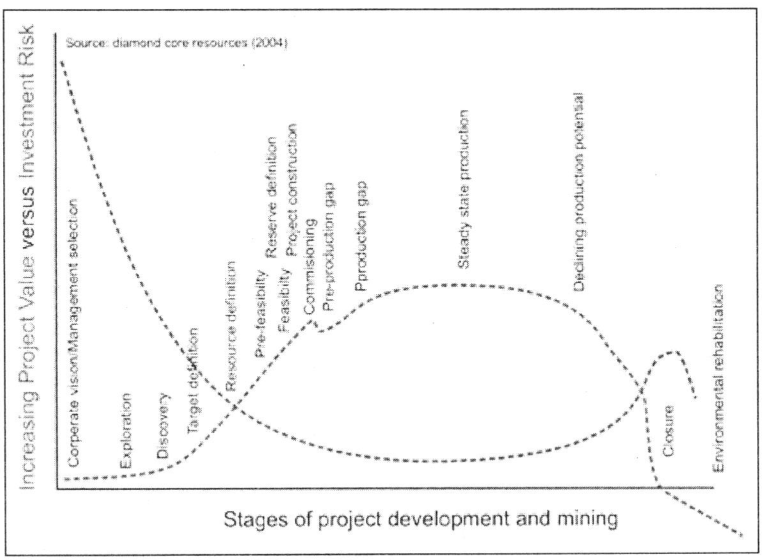

Application of appropriate methodology to value of mineral deposits is to aim to establish fair market value of mineral properties. According to I. S. Thompson, fair market value of exploration and underdeveloped properties can be determined by four factor:

1. geoscience factor,
2. cost,
3. market,
4. income.

Criteria and methods of mineral deposits evaluation were summarized below:

1. Net asset value (NAV) the net asset value stated in a company's report and account should theoretically be equal to its actual value. It is often not so in practice, especially when it comes to junior mining companies. At the early stages of exploration works this reflects solely what the company had already spent on its project. It could in some cases undervalue the company, if its project is successfully developed into an economically viable deposit. Or it could overstate its worth, since in most cases mineral projects will not be turned into a mine. Adopting a prospectivity enhancement multiplier (PEM) to historical expenditures in order to value to project more accurately remains very subjective, and is often questioned.
2. Ore in the ground, this type of valuation method could be described as a basic rule of thumb. It is used for valuations of different metallic projects when little real data is available. The gold industry uses it very commonly. For an exploration property analysts would use USD 30 to USD 50 per ounce of gold, moving up to USD 90 to USD 120 per ounce of gold when a company has proved that it can actually produce metal. Valuation of reserves is based on very generalised assumptions on mining, processing costs (which differ from one project to another) and operating margin (which moves with significant shifts in commodity prices).
3. Net Present Value (NPV), the most widespread method for resource companies. Theoretically, given the nature of the development timeline, it is best suited for prospecting companies. Typically the NPV method is applied with different discount rates (depending on the cost of capital, political risk, and project development risk) together with sensitivities analysis.
4. Other valuation approaches by industry buyers

Junior miners are usually being taken over by larger mining companies interested in their mineral assets for development of further production. Before they will go ahead with the process, they will have evaluated the junior, and the assets it is prospecting. For that purpose they use different valuation methods:

1. Prospectivity Enhancement Multiplier (PEM) of historical exploration expenditures, based on whether these works have enhanced the prospectivity of the ground or not. Historically PEM has usually ranged from 0.5 to 3.0 based on the subjective assessment of the acquirer.
2. Comparative Value Method – this method uses valuation multiples based on an average of "comparable" recent deals. It is used frequently by investment banks estimating the value of the project. However in early stage mining companies very little publicly disclosed data is available. So it is not always clear what key criteria have been most important in the final assessment, or whether a strategic premium has been paid.
3. Royalties or Farm-in Agreements – separate values are placed on both the expenditure that has already been committed and the future funding that the new entrant intends to provide. A discounted value of the latter points to the value that the buyer is placing upon its 'stake'.

NPV – it is more accurate when a potential acquirer is given access to internal company data in the due diligence process, than NPV valuation based on public data.

Option Pricing model – uses the option pricing model to calculate the commodity price at which the full risk adjusted NPV of the mine is greater than zero.

VALUATION OF JUNIOR MINING COMPANIES' MINERAL PROJECTS

In the recent years junior mining companies have become the most active players in the mining industry when it comes to exploration of mineral properties. The upward trend in metal prices has helped them raise capital for many mineral projects across the globe. Some of the juniors have been successful in selling their assets or being taken over by large mining houses that are typically interested in developing them to production stage. Whenever such a transaction is in place both sides will valuate a project based on their own views and methodical approach applied. Such transactions are usually done when a project is at prefeasibility or feasibility stage.

Junior mining companies tend to use the NPV method and relate this value to their quoted market value. If their market value is less than the NPV of the project they would most likely be reluctant to sell, hoping that their value will increase as the project is progressed further. They will also use other junior mining companies and projects as their benchmark. Such comparison is listed in the table below.

Table 1. Comparison of junior mining companies and their assets

Company	Project	Initial Capital	Annual Cu Production	Metal Prices	NPV	IRR	Stage	Market Capitalization
Peru Copper	Toromocho	$1.5 billion	273.000 tonnes	$1.10 Cu	$921 million (8%)	16.7%	Prefeas.	$826 million
Novagold	Galore Creek	$1.8 billion	119.000 tonnes	$1.50 Cu	$599 million (5%)	10.6%	Feas.	$1.65 billion
Northern Peru	Galeno	$976 million	144.000 tonnes	$1.35 Cu	$560 million (8%)	18.2%	Prefeas.	$311 million
Augusta Resources	Rosemont	$725 million	103.000 tonnes	$1.20 Cu	$442 million (8%)	17.0%	Scoping	$221 million
Northern Orion	Agua Rica	$2.1 billion	136.000 tonnes	$1.40 Cu	$372 million (8%)	12.0%	Feas.	$683 million
Chariot Resources	Marcona	$236 million	70.000 tonnes	$1.20 Cu	$364 million (8%)	28.0%	Scoping	$355 million
Equinox Minerals	Lumwana	$715 million	122.000 tonnes	$1.30 Cu	$321 million (8%)	15.0%	Feas.	$2.48 billion
Petaquilla Resources	Petaquilla	$1.7 billion	193.000 tonnes	$1.30 Cu	$287 million (8%)	13.3%	Feas.	$366 million
Corriente Resources	Mirador	$202 million	52.000 tonnes	$1.10 Cu	Pre tax $71 million	14.8% pre tax	Feas.	$374 million

Source: based on publicly available information as of Oct. 2007.

Senior mining companies interested in purchasing mineral projects owned by juniors listed on stock market will use slightly different approach. Rather than using the NPV method they will focus on the market value of a given junior, and look at comparable projects which had been transferred in the past. Apart from having to pay for the project itself they usually face bigger expenses related to mine construction. Thus they are not keen on the juniors' value to rise. However when commodities rally and the projects are scarce, they face a lot of competitors, which many a times leads to high prices of such transactions. Large mining companies often employ investment banks to evaluate and valuate potential acquisition targets for them. Those again pay a lot of attention to comparative value method. They will look at all precedent transactions regardless of their size and focus on the amount of money paid for a pound of metal or metal equivalent. They will take the average and set it as a benchmark for future deals. Below is a list of such transactions prepared by one of the leading Canadian investment banks.

Table 2. Comparative analysis of mining industry transactions

Asset	Date	Status	Sellers	Buyers	Development Stage	Metals	Percent Acquired	Total Value	Acquired Cu Tonnes	US$/lb Cu	Acquired Cu Eq Tonnes	US$/lb Cu Eq
Michiquillay	30-Apr-07	Completed	GOVERNMENT OF PERU	ANGLO AMERICAN PLC	FEA	CU AU AG	100%	$403.0	3,753,800	$0.049	4,992,711	$0.037
Tenke Fungurume	11-Apr-07	In progress	TENKE MINING CORP	LUNDIN MINING CORP	FEA	CU CO	25%	$1,300	3,446,710	$0.171	5,517,493	$0.107
Ruashi	26-Mar-07	In progress	SENTINELLE GLOBAL INVESTMENTS (PTY) LTD	METOREX LTD	PRD	CU CO	13%	$44.6	187,776	$0.108	439,594	$0.046
Tampakan	21-Mar-07	In progress	ALSONS CORP	INDOPHIL RESOURCES NL	FEA	CU AU	5%	$23.6	583,050	$0.018	718,099	$0.015
Cordillera de Las Minas	05-Mar-07	In progress	ANTOFAGASTA PLC, CVRD	PANORO MINERALS LTD	RD	CU AU	100%	$14.4	693,000	$0.009	873,000	$0.007
Rio Blanco	06-Feb-07	In progress	MONTERRICO METALS PLC	ZIJIN MINING GROUP CO LTD	FEA	CU MO	100%	$185.0	7,107,000	$0.012	8,911,000	$0.010
Red Chris	02-Feb-07	In progress	BC METALS CORP.	IMPERIAL METALS CORP	FEA	CU AU	100%	$71.3	1,810,573	$0.018	2,930,222	$0.011
Chibuluma South	02-Feb-07	In progress	INDUSTRIAL DEVELOPMENT CORP (GOV'T OF SA)	METOREX LTD	PRD	CU	39%	$21.0	71,750	$0.133	71,750	$0.133
Pebble	30-Jan-07	Completed	GALAHAD GOLD PLC	RIO TINTO PLC	RD	CU MO AU	10%	$79.5	1,902,719	$0.019	3,366,040	$0.011
Andocollo	05-Dec-06	Completed	COMPANIA MINERA DEL PACIFICO SA	AUR RESOURCES INC	PRD	CU AU	27%	$103.0	1,210,140	$0.039	1,435,537	$0.033
Vizcachitas	16-Nov-06	Completed	GLOBAL COPPER CORP	CHG RESOURCES LTD	RD	CU MO	79%	$11.5	631,560	$0.008	904,840	$0.006
Oyu Tolgoi	18-Oct-06	Completed	IVANHOE MINES LTD	RIO TINTO PLC	FEA	CU	20%	$691.0	3,506,982	$0.090	3,870,866	$0.081
El Tesoro	24-Aug-06	Completed	EQUATORIAL MINING	ANTOFAGASTA MINERALS	PRD	CU	39%	$401.0	379,860	$0.048	379,860	$0.048
Pebble	27-Jun-06	Completed	NORTHERN DYNASTY MINERALS LTD	KENNECOTT EXPLORATION INC	FEA	CU AU MO AG	10%	$77.9	1,828,827	$0.019	3,046,649	$0.012
Tintaya	16-May-06	Completed	BHP BILLITON LTD	XSTRATA PLC	PRD	CU AU AG	100%	$750.0	1,304,000	$0.261	1,371,666	$0.249
Tritton	02-May-06	Completed	TRITTON RESOURCES LTD	STRAITS RESOURCES LTD	PRD	CU AG AU	41%	$57.8	90,208	$0.291	96,614	$0.272
Ongopolo Mining	26-Apr-06	Completed	ONGOPOLO MINING AND PROCESSING CO	WEATHERLY INTERNATIONAL PLC	PRD	CU	97%	$91.0	460,080	$0.092	501,261	$0.083
Mount Milligan	19-Apr-06	Completed	GOLDCORP INC	TERRANE METALS LTD	RD	AU CU	100%	$106.0	542,000	$0.089	1,172,380	$0.041
Xietongmen	13-Apr-06	Completed	GREAT CHINA MINING INC	CONTINENTAL MINERALS CORP	RD	CU AU	50%	$53.0	322,000	$0.075	548,146	$0.044
Lumwana	22-Mar-06	Completed	EQUINOX MINERALS LTD	ZAMBIA CONSOLIDATED COPPER MINES LTD	FEA	CU AU CO	10%	$30.0	179,300	$0.076	179,300	$0.076
Regalito	14-Mar-06	Completed	REGALITO COPPER CORP	PAN PACIFIC COPPER CO LTD	RD	CU	100%	$137.0	3,241,000	$0.019	3,241,000	$0.019
Reko Diq	14-Feb-06	Completed	TETHYAN COPPER CO LTD; BHP BILLITON LTD	ANTOFAGASTA PLC;BARRICK GOLD	RD	CU AU PB ZN	75%	$200.0	9,263,250	$0.010	11,465,712	$0.008
Yubileinoye	13-Feb-06	Completed	BASHKORTOSTAN NATURAL RESOURCES AGENCY	URALS MINING AND METALLURGICAL CO	PRD	CU ZN AG AU	100%	$12.1	1,690,000	$0.003	2,238,658	$0.002
Adastra	18-Jan-06	Completed	ADASTRA MINERALS INC	FIRST QUANTUM MINERALS LTD	FEA	CU CO	100%	$248.3	1,376,660	$0.082	2,463,890	$0.046
Carlota	05-Oct-05	Completed	CAMBIOR INC	QUADRA MINING LTD	FEA	CU	100%	$37.5	355,000	$0.048	355,000	$0.048
Las Cruces	03-May-05	Completed	MK RESOURCES CO (LEUCADIA NATIONAL)	INMET MINING CORP	FEA	CU	70%	$88.5	739,900	$0.054	739,900	$0.054
									Min	$0.003		$0.002
									Max	$0.291		$0.272
									Average	$0.071		$0.058

Source: TD Securities, May 2007.

REFERENCES

A guide for Reporting Exploration and Information Mineral Resources and Mineral Reserves published in 1999 by The Society of Mining, Metallurgy and Exploration Inc.

Aussie mine 2008: Reaping the rewards - Review of trends in the Australian mid-tier mining industry, www.pwc.com, 20.12.2008

Corcoran K.: Junior Mining Investor, November 2007

Junior mine 2008: review of the trends of the AIM mining industry, www.pwc.com 10.01.2009

Kicki J. Nieć M. 2006: On the road to standardize mineral resources/reserves classification, Mineral Resource Management, Kraków 2006, Quarterly Volume 22 special Issue 2.

Steinitz J. 2008: How to value mining share, www.resourceinvestor.com/pebble.asp?relid=26537 10.12.2008

www.tdsecurities.com

www.srk.com

International Mining Forum 2009, Sobczyk, Kicki & Saługa (eds) © 2009 Taylor & Francis Group, London, UK, ISBN 978-0-415-80428-8

The Optimum Lifetime of a Polish Hard Coal Project Using Discounted Cash Flow Analysis

Piotr Saluga
Mineral & Energy Economy Research Institute
of the Polish Academy of Sciences, Krakow, Poland

ABSTRACT: The Polish hard coal industry produces over 90 Mill. tonnes yearly. The output comes overwhelmly from longwall systems. Such mining method is capital-intensive therefore one has been looking for technical, economic and time factors that will maximize a net present value (NPV) of a typical underground project. One of these factors is the lifetime of the project, which – for the sake of time value – is fundamental. The paper, making Polish-specific assumptions, gives the optimum rate of extraction and, consequently, the lifetime for the typical hard coal project, that would maximize the project's NPV.

KEYWORDS: Hard coal, discounted cash flow, DCF, net present value, NPV, project lifetime, optimum lifetime

1. INTRODUCTION

For the last decade in Polish hard coal mines has been continued a dynamic increase in longwall output. It is caused by constant modernization of extraction process – especially by improving both machinery and working system.

Table 1 presents the daily production in Polish mines in relation to seam thickness. The typical output in most Polish hard coal longwalls amounts to 3,000 tonnes/day but there are still 31% of longwalls producing less than 2,000 tonnes/d. Most (about 75%) of the output comes from 1.5–3.5 m thick seams [1].

Table 1. Number of longwalls in groups of daily production in Polish mines [1]

#	Seam thickness	Daily output [t/d] < 2,000	2,000–3,000	3,000–4,000	> 4,000	Total	Longwall output [t/d]
		Number of longwalls					
1.	< 1.5	3	0	0	0	3	1,221
2.	1.5–2.0	13	6	7	1	27	2,266
3.	2.01–3.5	18	20	21	9	68	3,014
4.	3.51–4.5	4	2	3	3	12	3,100
5.	> 4.5	3	6	4	3	16	3,125
Total		41	34	35	16	126	2,833

The reasons for increasing in output are mostly economic. The crucial factor for economic extraction is a face advance which results in lifetimes of hard coal projects. Unfortunately, in Polish conditions an average production rate (ca. 3,000 tonnes/d) is in many cases insufficient for economic effective extraction. Thus, the operating lifetime of a mining project is among the most important factors influencing the ultimate economic success of the project. The paper presents calculations of the optimal lifetime, searching appropriate production face advance rate, for a typical Polish hard coal project using discounted cash flow, DCF, analysis.

The optimal lifetime of the project can be defined as that period at which the net present value, NPV, (or internal rate of return, IRR) is maximized. In our case the NPV value was maximized.

The profitability of the project depends on:
1) the tonnage of reserves,
2) capital and operating costs,
3) the market prices (depending on demand and the quality of a product),
4) production rate,
5) process recovery,
6) project risks.

2. A DISCOUNTED CASH FLOW MODEL

Under the discounted cash flow model, the value of a project consists of the value of future cash flows (or other net revenues), less initial investments.

"Cash flow" is an accounting term, representing the balance of all cash revenues minus cash operating and current capital investments, as of the end of the fiscal year. It represents the volume of monetary after-tax assets in cash earned or lost over the year. Cash flows can be calculated ether on a constant or inflated currency basis.

Since investors would prefer to obtain benefits from a project as soon as possible, the value of each annual cash flow generated during project realization must be corrected with respect to the time value of money. The current (up-to-date) cash flow value (present value, PV) obtained after time t is calculated by discounting its future value (FV) according to the appropriate discount rate.

The basic parameters used in discounted cash flow analysis in order to measure the profitability of an investment are net present value (NPV) and internal rate of return (IRR).

NPV, which provides the basis for the appraisal of investment projects, consists of the sum of the present values of all the annual cash flows, minus the initial investment. NPV reflects the project value at a given discount rate, given a range of assumptions regarding the cash flows; it is thus a measure of the investment's worth. The equation used to calculate NPV is as follows:

$$\mathrm{NPV} = \left[\sum_{t=1}^{n} \frac{\mathrm{CF}_t}{(1+i)^n} \right] - I_0 \tag{1}$$

where: CF_t is cash flow in year t; I_0 is initial investment; i is discount rate, and n is total number of years required to realize the project.

An increase in the discount rate causes a decrease in the NPV.

Internal rate of return (IRR) is defined as that discount rate at which NPV equals zero. On the other hand, IRR can also be defined as the rate, which makes the cost of the initial investment equal to the present value of the future cash flows. The higher IRR, the more profitable the project, in the sense of return on the invested capital. The difference between the discount rate and IRR consists in the fact that the former is chosen by investor while the latter is determined by the characteristics of the cash flows. It can thus be stated that IRR is specified internally (determined in essence by the character and nature of the project), while the discount rate, which is indispensable in order to calculate NPV, is specified externally.

Although in principle NPV and the maximization of the company's value are theoretically sufficient criteria for ranking investments, NPV does not show the return we can expect from the money invested. This disadvantage may be compensated using IRR, which explains why the latter is the most popular criterion for assessing investment projects in the mining industry [2].

Most manuals state however that NPV is preferred merit measure, both for theoretical and supposedly practical reasons [2]. Also, NPV has seemed to be more adequate measure and therefore it was taken for the purpose of project evaluation discussed here.

3. ASSUMPTIONS FOR AN ECONOMIC VIABILITY ASSESSMENT

Calculations were made for a typical project (TP) conducted in the "X" hard coal mine. The mining panel of the TP project encompasses 8.0 million tonnes, occurring in a regular, continuous seam. The geometric parameters of a typical longwall in the TP are as follows: length – 250 m, height – 2.2 m, panel length – 1,500 m. A typical face advance in the "X" mine is relatively high – 5 m/d which results in 3,795 tonnes/d.

The following assumptions for the base case of the TP project were made:

3.1. *Development schedule*

A development schedule of the project TP will take 18 months. For the reasons of time value, development works were assumed to start in January of the year "0" and finish in June of the year "2". The development will be doing with 167.0 m/month.

3.2. *Capital expenditures*

The required capital expenditures were taken as covering for:
- drivage, support and equipping of development workings,
- longwall face machinery and equipment.

The development of the TP panel requires about 3 km of tunnels. It was assumed, that 80% of tunnels were to driven in host rock and 20% in coal. The average cost of drivage at mine "X" is ca. 7,500 Polish zlotys/m (3,125 US$/m), which means capital development costs of ca. 22.5 million zlotys (~9.4 million US$ – Table 2).

Table 2. Development workings to the TP panel at the "X" mine

Item	Length (assumed) [m]	Cross-sectional area (average) [m^2]	Production [thousand tonnes]	Drivage time [months]	Cost of drivage [million PLNs*]
Tunnels in rock	2,700	18.0	7.45 (stone)		
Tunnels in coal	300	18.0	97.2		
TOTAL tunnels	3,000	18.0	104.7	13.7	20.4

*PLN – Polish zloty ~ 2.4 US$.

The longwall was to be 250.0 m long, hence its support would comprise 166 units. Overall expenditure required to equip a standard longwall face with supports, a longwall shearer, scraper conveyors: face and end-gate, and a crusher amounts to ca. 70.2 million zlotys (~29.3 million US$). Investment cost expenditures schedule were put together in Table 3.

Table 3. Investment cost expenditures schedule in the TP panel

	Year "0"	Year "1"	TOTAL
Development workings [PLNs]	14,987,250	7,512,750	22,500,000
Face equipment [PLNs]		70,200,000	70,200,000
Investment cost TOTAL [PLNs]	14,987,250	77,712,750	92,700,000

*PLN – Polish zloty ~ 2.4 US$.

3.3. *Depreciation*

For the purpose of the calculations it was assumed that depreciation deductions would start – in the case of capital development workings – at the moment the development was completed, and in the case of the longwall equipment – at the moment the face was fully equipped. Annual depreciation rates were assumed as follows:

- longwall shearer, scraper conveyors – 25.0%,
- longwall support units, crusher – 20.0%,
- access drives – 4.5%.

It was further assumed that after longwall equipment was fully depreciated before the panel in question was mined out, the capital costs will be incurred renewal expenditures. The end value of the capital drives at the moment the panel had been exhausted was taken into account in cash flow calculations.

3.4. *Other assumptions*

The other assumptions were made as follows:

- Coal sale price: 180.0 zloty/tonne (~75 US$ /tonne).
- Operating cost: 160.0 zloty/tonne (~66.7 US$/tonne).
- Production losses: 5.0%.
- Working capital: 30 million PLNs (ca. 12.5 million US$).
- Process recovery: 90.0%.
- Waste location cost: 15.0 zloty/tonne (~6.3 US$/tonne).
- Inflation rate: 3.5%.
- Risk-adjusted discount rate, RADR (real): 8.0%.

Discount rate, RADR, was taken from Smith [3]. The calculations were done in current money using a nominal discount rate calculated as 11.8%.

4. ANALYSIS PROCEDURE

At first a Microsoft Excel DCF spreadsheet calculating NPV values was constructed (Table 4). Of course, the greater a face advance (and production rate), the greater NPV values. But higher rates require greater capital costs. On the other hand, greater capital engaged causes lowering of operating costs. The management of "X" mine had developed estimates of capital and operating costs depending on various advance rates (Figure 1). Costs and revenues were estimated at various production rates for the project and the spreadsheet was repeatedly calibrated at various production rates and costs for the project. The process was continued in an iterative fashion until the highest NPV has been determined. The obtained production rate is the optimum one. It determines a project optimum lifetime.

Table 4. Cash flow spreadsheet for the TP project of the "X" hard coal mine

Process recovery [%]:	90
Longwall reserves [tonnes]:	7 807 490
Panel reserves [tonnes]:	8 000 000
Production losses [%]:	5
Average mining hight [m]:	2,20
Production rate [tonnes/month]	79 695
Lifetime of mining [months]:	100,4
Discount rate RADR (real) [%]:	8,0
Inflation rate [%]	3,5
Discount rate RADR (nominal) [%]	11,8

Operating cost [PLNs/tonne]:		160,00
Capital investment [PLNs]:		162 900 000

	Okres:	year 0	year 1	year 2	year 3	year 4	year 5	year 6	year 7	year 8	year 9
Output [tonnes/year]		4 964	480 658	956 340	963 792	956 340	956 340	956 340	956 340	956 340	827 450
Waste rock [tonnes]		65 241	80 521	95 634	193 579	95 634	95 634	95 634	95 634	95 634	82 745
Coal production [tonnes]		4 467	432 592	860 706	867 413	860 706	860 706	860 706	860 706	860 706	744 705
Coal price [PLNs/tonne]		180	180	180	180	180	180	180	180	180	180
Waste location cost [PLNs/tonne]		15	15	15	15	15	15	15	15	15	15
Capital investment [PLNs]		14 987 250	77 712 750	0	0	0	18 840 000	51 360 000	0	0	0
face equipment		0	70 200 000	0	0	0	18 840 000	51 360 000	0	0	0
development workings		14 987 250	7 512 750	0							
Working capital [PLNs]		30 000 000									
Gross earnings [PLNs]		804 132	77 866 632	154 927 080	156 134 304	154 927 080	154 927 080	154 927 080	154 927 080	154 927 080	134 046 900
minus: Waste loction costs [PLNs]		978 619	1 207 814	1 434 510	2 903 688	1 434 510	1 434 510	1 434 510	1 434 510	1 434 510	1 241 175
minus: Operating costs [PLNs]		647 773	62 725 898	124 802 370	125 774 856	124 802 370	124 802 370	124 802 370	124 802 370	124 802 370	107 982 225
Gross profit [PLNs]		-822 260	13 932 920	28 690 200	27 455 760	28 690 200	28 690 200	28 690 200	28 690 200	28 690 200	24 823 500
minus: Depreciation [PLNs]		0	7 997 250	15 994 500	15 994 500	15 994 500	15 994 500	15 994 500	15 994 500	15 994 500	12 118 640
Taxable income [zł]		-822 260	5 935 670	12 695 700	11 461 260	12 695 700	12 695 700	12 695 700	12 695 700	12 695 700	12 704 860
minus: Taxes [PLNs]		0	1 127 777	2 412 183	2 177 639	2 412 183	2 412 183	2 412 183	2 412 183	2 412 183	2 413 923
Net income [PLNs]		-822 260	4 807 893	10 283 517	9 283 621	10 283 517	10 283 517	10 283 517	10 283 517	10 283 517	10 290 937
plus: Depreciation [PLNs]		0	7 997 250	15 994 500	15 994 500	15 994 500	15 994 500	15 994 500	15 994 500	15 994 500	12 118 640
minus: Capital investment [PLNs]		14 987 250	77 712 750	0	0	0	18 840 000	51 360 000	0	0	-30 822 610
minus: Working capital [PLNs]		30 000 000	0	0	0	0	0	0	0	0	-30 000 000
Net cash flow [PLNs]		-45 809 510	-64 907 607	26 278 017	25 278 121	26 278 017	7 438 017	-25 081 983	26 278 017	26 278 017	83 232 187
Discounted cash flow [PLNs]		-45 809 510	-58 067 281	21 031 216	18 098 912	16 832 017	4 262 228	-12 858 113	12 051 576	10 781 514	30 550 221
NPV [PLNs]	-3 127 221										

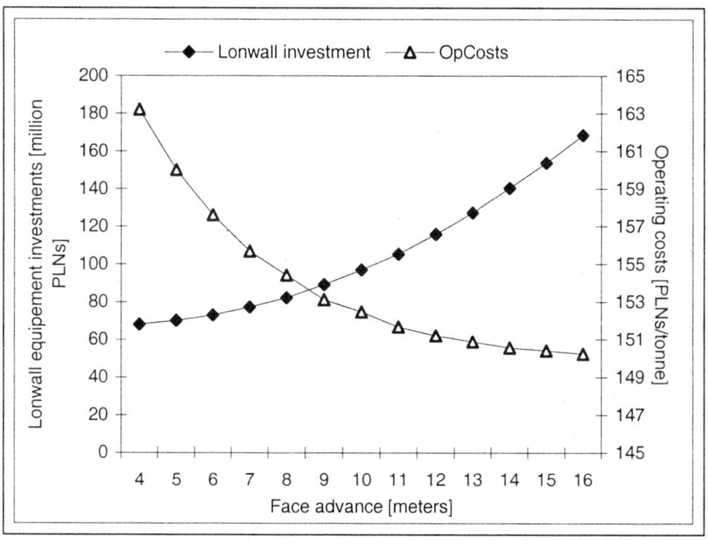

Figure 1. Capital and operating costs for hard coal mine "X" depending on various face advance parameters (self study)

5. RESULTS OF THE ANALYSIS

Results of the analysis are presented in Table 5 and Figure 2. NPV for the base case (face advance 5 m/d) is negative: ca. −3.1 million zlotys (−1.3 million US$) and the venture is uneconomic. Greater face advances and, in result, greater production rates cause higher NPV values. This is mainly because time value of money. There are drastic changes in NPV value with first changes of an advance: an increase from 5 to 6 m/d (3,795 to 4,554 tonnes/d) results in a large increase in NPV – up to 20 million. Next changes are also powerful – increment in advance from 6 to 8 m/d (up to 6,072 tonnes/d) causes successive 20 million increase in NPV. The highest NPV occurs at a production rate 9,867 tonnes/d (13 m/d). This is, then, the optimum production rate for the "X" mine. Thus, the optimum life of the TP project is 56.6 months (4.7 years).

Table 5. Face advances, project lifetimes and resulting NPVs for the "X" mine

Face advance [m/d]	Project lifetime [years]	NPV [million PLNs]
4	12,0	–29,8
5	9,9	–3,1
6	8,5	20,0
7	7,5	40,2
8	6,7	58,9
9	6,1	66,7
10	5,7	77,0
11	5,3	81,4
12	5,0	88,9
13	4,7	89,3
14	4,5	88,7
15	4,3	87,4
16	4,1	85,6

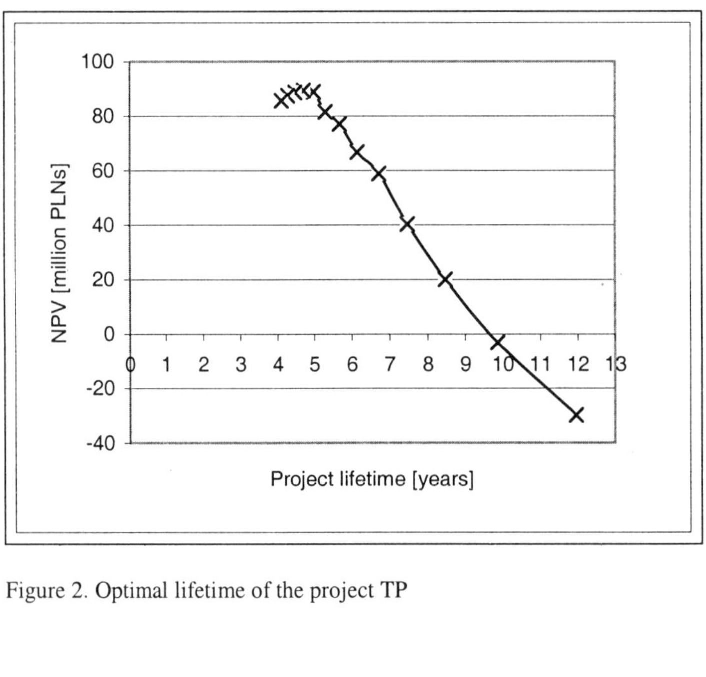

Figure 2. Optimal lifetime of the project TP

CONLUSIONS

After a sector restructuring in the past decade, Polish hard coal mining with a yearly output of ca. 90 million tonnes is still an important branch in Poland. Unfortunately, in spite of several reforms and improvements, Polish hard coal mining industry has been still balancing at the edge of productivity. At present, the best Polish hard coal mines have concentrated at increasing in concentration of longwall output. It seems that the answer to profitable mining lies in selecting an appropriate production face advance rate. Obtaining higher rates of extraction results, however, in greater investment costs. On the other hand one can expect decreasing in operating costs. This paper, presenting an example of a mining project conducted in "typical" Polish mine, gives an idea of conditions for optimal extraction. Respecting various assumptions the optimum production rate was calculated to 9,867 tonnes/d (13 m/d). High rates of advance come with a number of consequences, linked mainly to the creation of high stresses in the rockmass. In favourable mining and geologic conditions high stress provides for easier breaking of coal at the face and improved management of the hangingwall and the supports. It is advantageous in the case of weak roof and foot rocks, although in the latter case high face advance may not be recommended as increased stresses cause footwall heave. Footwall movement, even at a limited scale, may be critical due to low height of the face. The greatest threat, however, is associated with methane hazard – high rates of mining result in high volumes of gas being released into the mine atmosphere. To create high stresses may prove dangerous in the case of coal seams prone to seismicity and where the hazard of rock-and-gas ejections is present [4]. In the case of coals prone to spontaneous combustion high rate of mining magnifies this danger. Any mine planning should therefore to face all the mentioned problems, taking following ventures:

1. ensuring continuous mining with keeping possible steady rate of advance;
2. using of preventive actions against methane, coal dust or fire hazards;
3. undertaking of adequate actions against seismic hazard.

REFERENCES

[1] Sikora W. 2008: Scenariusz rozwoju technologii wydobycia węgla w warunkach utrudnień górniczych. (In Polish). Wirtualny Przemysł Górniczy.

[2] Torries T.F. 1998: Evaluating Mineral Projects: Applications and Misconceptions. Society for Mining, Metallurgy and Exploration, Inc.

[3] Smith L.D. 2000: Discounted Cash Flow Analysis. Methodology and Discount Rates. Special Session on Valuation of Mineral Properties – Mining Millennium 2000, March 8, Toronto.

[4] Bodarenko V., Kuzmenko O., Dychkowsky R. 2004: Prospects of Development of Highly Productive Coal Extraction Technologies. Proceedings of the 5th International Mining Forum 2004, A.A. Balkema, Leiden, London, New York, Philadelphia, Singapore.

International Mining Forum 2009, Sobczyk, Kicki & Saługa (eds) © 2009 Taylor & Francis Group, London, UK, ISBN 978-0-415-80428-8

Managing Net Working Capital at a Mining Company Operating As a Multi-Facility Enterprise

Maria Sierpińska, Arkadiusz Kustra
AGH – University of Science and Technology, Krakow

Łukasz Siodłak
Ph.D. student at the Faculty of Mining and Geoengineering, AGH – University of Science and Technology, Krakow

INTRODUCTION

Net working capital plays a significant role in a company as maintaining it determines the risk con nected with financing current assets. This risk is usually due to the possible difficulties with selling the manufactured products or purchased goods and collecting accounts receivable for their sales. This capital allows financial liquidity to be maintained and changes in its level impact the current situation of the company specifically with regard to liquidity. The basic premise for winding up a company is its loss of liquidity, maintaining which depends on the company possessing the appropriate level of working capital. Consequently, it is very important to manage working capital on a current basis, constantly track changes in its level and the demand for it. This is because for many companies, net working capital constitutes a principal element of their financial resources.

The analysis of working capital and synthetic practical conclusions presented in this article were formulated based on research at a hard coal mine operating as part of a multi-facility enterprise. The role of the company organisational structure as well as the resultant authorities and autonomy in developing the financial policy are presented in the light of managing net working capital.

1. THE ROLE OF NET WORKING CAPITAL IN MANAGING THE FINANCIAL BALANCE OF THE COMPANY

Net working capital is easiest defined as the difference between current assets (working assets) and current liabilities (short-term liabilities). It is also defined, e.g. by M. Sierpińska[1], as: the part of assets which is not financed with current payables, but with long-term capital, or as the difference between the working assets and short-term capital, and it is equivalent to the part of working assets financed with equity and its derivatives as well as long-term liabilities. D. Wędzki also quotes the definition of net working assets laid down in the International Accounting Standards (IAS 1, 55): "If a unit supplies products or provides services within a clearly defined operating cycle, the split in the balance sheet into short- and long-term assets and payables provides useful information by distinguishing those net assets which are in constant trading, as working capital, from those that the unit uses in the long term".[2]

[1]M. Sierpińska, D. Wędzki: Zarządzanie płynnością finansową przedsiębiorstwa. Wydawnictwo Naukowe PWN, Warszawa 2002, p. 74.
[2]D. Wędzki: Strategie płynności finansowej przedsiębiorstwa. Oficyna Ekonomiczna, Kraków 2003, p. 39.

The level of net working capital may increase or decrease as a result of subsequent business operations taking place in the company, and so:

1. Working capital increases:
   - as a result of decreasing the balance of fixed assets by monetizing superfluous assets, liquidating worn-out assets and reducing investments;
   - as a result of increasing long-term funds (allocating the profit earned for development), increasing equity (issuing shares) and increasing long-term liabilities (bank loans, bonds).
2. Working capital decreases:
   - as a result of an increase in fixes assets caused by purchases of intangible assets, tangible assets, shares and bonds, extending long-term loans to other companies or increasing the level of long-term receivables;
   - as a result of reducing long-term funds (redeeming shares), covering losses with supplementary capital, repaying long-term liabilities.

Movements in net working assets, whether up or down, always reflect the current situation of the company with regard to maintaining its liquidity. When the level of working capital is too low, this means that short term payables cover a larger proportion of working assets. Too high short-term payables, in turn, may cause liquidity to be lost. If the level of working capital is too high, this means that long-term funds cover a larger proportion of working assets. If the source for raising the working capital is a loan or an advance, the company incurs financial costs which reduce its financial result, but if it is a raise of the equity, the consequence is that the rate of return on equity for the shareholders drops. In that case, the company has no problem to pay its current payables on time, but it may experience over-liquidity caused by funds being frozen in working assets, mainly inventory (stocks) and accounts receivable.

The following functions are generally attributed to net working capital[3]:

- it is a measure of liquidity as it shows the amount of assets which secure the repayment of mature payables (the higher the capital, the lower the risk of losing liquidity);
- it is a buffer protecting the company from operating losses, as maintaining a surplus of working assets over short-term payables allows the operating cycle to renew at the assumed level (the more working capital, the better this protection);
- it allows the negative impact of the environment – such as changes in demand, delayed deliveries, trading partners not paying their dues – on the company to be reduced.

2. NET WORKING CAPITAL AND THE DEMAND FOR NET WORKING CAPITAL

Another notion closely connected with the notion of net working capital as a part of long-term funds financing the working assets of a company is the notion of the demand for working capital. The demand for working capital is due to the difference between the value of working assets held and the value of short-term payables arising in connection with the business conducted.[4]

The value of the demand for working capital is impacted directly by, *inter alia*:

- the inventory (stock) of materials and raw materials for production, goods, production in progress and finished products;
- the value of receivables for deliveries and services;
- the value of cash reserves;
- the value of payables for deliveries and services, for taxes and similar charges, for salaries.

---

[3] L. Bernstein: Analysis of Financial Statement. Homewood: Irwin, 1993, p. 87, quoted following D. Wędzki: op. cit. p. 40.

[4] A. Kusak: Płynność finansowa, Analiza i sterowanie, Wydawnictwo Naukowe Wydziału Zarządzania Uniwersytetu Warszawskiego, 2006. p. 71.

The increase in the value of inventories, short-term receivables and cash reserves raises the demand for working capital, while a decrease leads to reducing that demand. And the other way around: an increase in short-term payables leads to a fall in the demand for capital, while a decrease – to a growth. For example, the following will boost the demand for working capital: trading partners receiving longer payment terms, which leads to an increase in receivables, or reducing the time for paying dues to suppliers, or an increase in sales leading to a growth in inventories and receivables.[5]

A comparison of the level of net working capital to the demand for that capital may reveal the following demand situations - the demand may be:

- equal to net working capital;
- greater than net working capital;
- lower than net working capital.

Regardless whether there is a shortage or a surplus of net working capital, the company should take action aimed at restoring the balance between working capital and the demand for it, as a shortage may expose the company to problems with financial liquidity and eliminate a part of the profit, while a surplus causes superfluous costs for the business, thus hurting is market position.

A company can use the following methods to restore the balance between net working capital and the demand for it[6]:

- reduce the demand for working capital by cutting the level of current assets or raising short-term sources of finance, for example by: optimising inventories, offering buyers rebates for cash payments or payments before the maturity date, taking out short-term loans, negotiating longer terms for paying liabilities to suppliers;

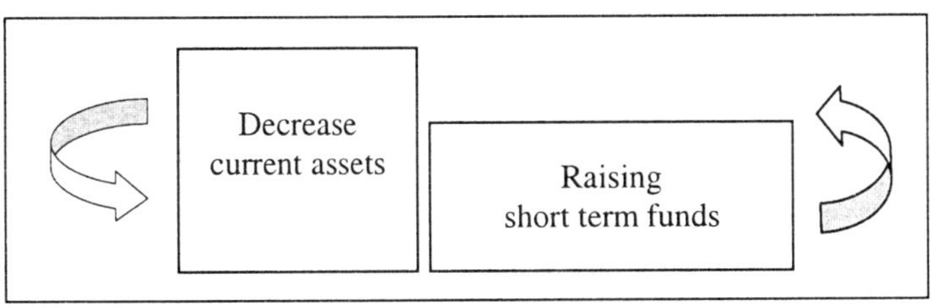

- increase net working capital by reducing fixed assets or raising long-term funds, for example by: selling the financial assets held, selling separate tangible fixed assets, retaining the profit earned, taking long-term loans, raising equity by issuing a new series of shares;

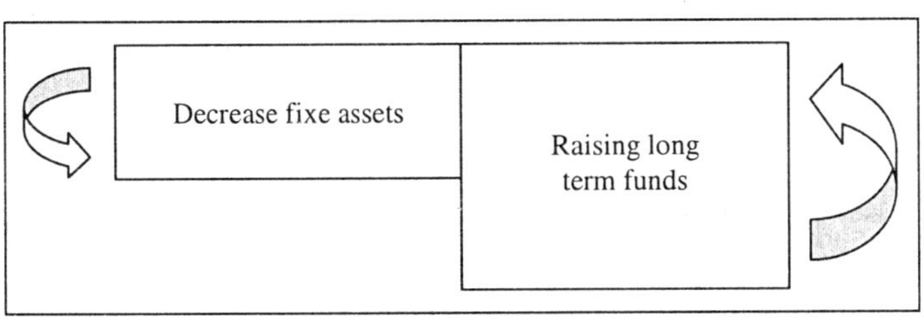

- use a mixed solution, called the financial mix, by applying and correctly choosing actions from the first and the second methods.

[5]M. Sierpińska, D. Wędzki: op. cit. p. 89.

[6]M. Sierpińska, T. Jachna: Metody podejmowania decyzji finansowych. Wydawnictwo Naukowe PWN, Warszawa 2007, p. 126–127.

The above considerations show that there is a major link between net working capital available to a company and the demand for it. Changes in the demand for working capital make it necessary to constantly and thoroughly watch its level so as not to allow liquidity to be lost.[7] So the management of every company should pay a lot of attention to the subject of net working capital and use it for the current management of liquidity.

3. DETERMINANTS OF NET WORKING CAPITAL MANAGEMENT IN MINING COMPANIES

In the case of mining companies, an analysis of working capital and its management must take into account the specific nature of production, the technology used as well as external factors caused by seasonal sales and organisational structures found in mining.

From the resource perspective, the analysis of the demand for capital applies in particular to the strategy of managing inventories, receivables and payables to suppliers in the form of mercantile credit.

In mining companies, the inventory depends on the volume of production and sales, and therefore on the utilisation of tangible fixed assets. It is well known that mining companies generally have significant fixed assets which bring about certain maintenance costs. As a rule, the high level of fixed costs associated with these assets causes low flexibility in operations, and therefore a poor ability to adjust to changing market conditions. Companies try to reduce the high level of operating risk due to the asset structure by utilising the assets to the maximum and thus reducing total unit costs.

As a result, in an attempt to utilise their assets and minimise the cost of idle production capacity, mining companies have to generate a high inventory of extracted but unsold raw material. The cost of maintaining the inventory is due to freezing capital in those items, which is usually considered to be an opportunity cost, and incurring storage costs, usually identified as simple costs expressed as types. Mining companies try to rationalise their inventory level by building a sale network of representatives or dealers who can take over the function of surplus storage. In some cases, the storage cost can also be limited by shifting it to the buyer under a contract providing for deposited stocks at the trading partner's premises, who pays the producer – the mine – only after selling the raw material.

Apart from the inventory, an important part of maintaining liquidity of mining companies is the strategy of managing receivables from customers. The credit policy of the company is usually assessed by analysing the payment terms, rebates granted, debt in the form of mercantile credit, the penalty interest and the credit risk of trading partners. The ability to use certain tools depends on the competitive position of the mining company against the customer. Practice shows that mines most frequently use long-term contracts which reduce operating risk by securing defined sales at the contractual price. Prices can be negotiated at specified intervals, and, as a rule, prices under such contracts are different from spot prices: when there is an upturn in the commodity market, fixed contractual prices significantly reduce the revenue of mining companies. The payment terms offered and the maximum debt of trading partners in the form of mercantile credit are due to the relationship with the customer so far. This means that an aggressive mercantile credit policy (long payment terms in trading) can be applied to a customer with whom the company has a trading history showing the customer to be reliable and of low credit risk. Otherwise, the policy of debt management is conservative.

---

[7] W. Wawrzyńczak: Zarządzanie kapitałem obrotowym w przedsiębiorstwie. Zarządzanie finansami firm – teoria i praktyka. Tom 2. Ed. W. Pluta, Wrocław: Wydawnictwo Akademii Ekonomicznej im. Oskara Langego 2003, p. 282.

The strategy of inventory or receivables management must also take into account the market conditions and trends to which mining companies are subject. It is well known that there are business cycles in the mining market, and they determine prices and therefore also the behaviour of companies. If there is a booming market for minerals, the policy of managing operating assets can be aggressive, which coupled with the simultaneous operating risk of mining companies causes disproportionate changes of financial results. World-wide practice shows that to reduce the risk of changing prices of minerals, mining companies run an expansive policy of product and geographic diversification.

The level of inventory and receivables of the company generates a specific demand for net working capital. A part of current operations is paid for with mercantile credit from the suppliers. Mercantile credit is the cheapest source for financing current operations, as using it causes no financial costs. In the financial perspective, payables to suppliers may lead to costs if they become overdue, and a supplier with a strong competitive position is not afraid to charge penalty interest. As a rule, mining companies impose the principles of liability payment to suppliers, who accept extended payment terms not to lose their selling market. This way, mining companies finance their operations with mercantile credit, which replaces bank loans in their sources of finance. Many mining companies use no external sources of finance generating an interest expense, as they put a greater value on financial security and reducing financial risk then on the higher efficiency of using capital, particularly equity. This leads to a situation in which the demand for net working capital due to the strategy of current asset management is covered from long-term sources, in particular equity.

A safe policy of financing current operations from long-term sources in the form of net working capital is contrary to the interests of the shareholder, who wants to use the equity efficiently. If the equity is used inefficiently, the shareholder will try to generate profits by paying out dividend, and thus will reduce the long-term capital employed in operations. When current operations expand and bring about significant inventories and receivables, obtaining long-term capital becomes the source of supply of net working capital which must balance the demand for that capital.

An analysis of capital and the demand for it in the dynamic approach focuses on the ability to generate cash flows. What is particularly important are operating cash flows, in which the company recognises sources of income and expenditure connected with current operations. A positive balance of cash from operations proves that the enterprise is profitable in cash terms. This surplus may finance the remaining areas, namely the investment and financial activities, if cash shortages occur there.

When one analyses financial flows, their main source is the profit of the company which is the difference between all its revenues and costs. However, the presentation of financial result in accruals accounting is not the only factor that determines liquidity. What is more, practice shows that the majority of companies were profitable at the moment they went bankrupt, but the results achieved were not reflected by real cash flows. So apart from the bottom line, an important issue concerning cash sources is the depreciation of tangible assets and the amortization of intangibles, which, although they constitute costs reducing profitability, are not expenses. So the value of assets, the asset management policy applied and the depreciation/amortization methods adopted determine the dynamic current liquidity of a company. In practice, when companies present their financial figures, they attach results based on such measures as the EBITDA or the financial surplus (net profit + + depreciation/amortization), at the same time reporting the adjustment of accounting result to account for depreciation/amortization.

Just as depreciation/amortization, provisions set up can also form a source of cash flows. They constitute costs of the company, but they also represent money held for future expenses connected with the business conducted. Those may be provisions for social performances connected with the social security of employees and other provisions. In the case of mining companies, the closure fund established under the mining and geology law forms an important part of provisions. If its estimate does not correspond to the potential expenditure for closure, an expert auditor recommends that additional provisions for the closure and remediation be set up. These are tax deductible expenses under tax accounting, but from the practical point of view, a mining company keeps this money on separate bank accounts or in treasury securities, i.e. risk-free assets. If some money from the set-up closure provisions is spent, this part of provisions is no longer a tax-deductible expense.

Another significant issue concerning operating cash flows of a company is reducing the cash taken by the demand for net working capital. This translates into restructuring the inventories and accelerating their turnover compared to revenues, reducing the exposure due to mercantile credit extended to buyers and simultaneously boosting sources of finance which are spontaneous in nature and generate no interest expense when used. What is also important in the capital demand is an analysis of movements in accruals and prepayments, which have the nature of provisions for operations.

4. MANAGING NET WORKING CAPITAL AT A HARD COAL MINE

The shareholder restricts the business of the analysed mine with regard to the autonomy and independence in the management of assets and financial resources. In the assets area, the mine is responsible for managing current assets, namely inventories and receivables, but has no independence in developing the investment policy for managing fixed assets. On the sources of finance side, the analysed mine uses the equity provided by the shareholder and the generated profits, while external capital is dominated by spontaneous, interest-free liabilities. So the mine does not use loans and advances that would determine financial costs.

The analysis of net working capital covered 12 months of 2007 and 9 months of 2008.

In 2007, fixed assets were seen to grow by 33%, and working assets by 57%. At the same time, working assets represented only 16.6% of total assets at the end of 2007, which means that the rest consisted of fixed assets, particularly tangible ones. This confirms structures that are characteristic for mining companies operating in the heavy industry, where the asset structure includes a lot of assets that are not mobile and generate fixed costs due to maintaining the production capacity. At the same time, it should be noted that the above asset structure leads to a high operating risk combined with poor flexibility in operating and reacting to changing market conditions.

On the side of sources of finance, at the beginning of 2007, the mine followed a conservative liabilities management policy, where long-term capital accounted for over 60%, and short-term capital for 40%. In late 2007, the structure of liabilities changed, and the strategy of managing sources of finance became aggressive. This meant that the share of long-term capital in the liabilities dropped to 36%, while short-term capital represented 64% of the total resources held.

An analysis of horizontal links between assets and liabilities as part of analysing net working capital showed it to be negative throughout 2007, and as a result of a significant increase in short-term liabilities compared to current assets, this negative value even increased. The growth of liabilities was due to spontaneous liabilities arising as part of standard business operations, not liabilities generating financial expense for the mine. In the analysed period, the mine did not use external sources of finance accruing interest, like loans or advances. The negative net working capital proves that short-term sources are financing not only all current assets but also fixed assets, and therefore investment activities. Only in 2007, the utilisation of short-term sources for investment operations rose from PLN 117,917 thousand to PLN 296,496 thousand (Table 2). The reasons for movements in net working capital are shown in the fund flows for 2007 (Table 3) in which one can see a significant drop in long-term funds (in June) and a growth in fixed assets due to purchasing tangible assets (in October), while the financing gap was covered with short-term liabilities in the form of mercantile credit.

In the 9 months of 2008 that were analysed, resources (total assets, total liabilities) of the mine under consideration shrunk by some 8%. The percentage breakdown of assets was as follows: 12%

of working assets, 88% of fixed assets. On the side of sources of finance, short-term payables (specifically to suppliers) were cut from PLN 400,122 thousand at the beginning of 2008 to PLN 292,789 thousand at the end of September 2008. Long-term capital, to which profit was credited, grew from PLN 222,666 thousand as of early 2008 to PLN 281,186 thousand as of the end of September 2008.

Regardless of the changes reported in 2008 in the balance sheet structure, net working capital remained negative, and therefore the utilisation of short-term payables in investment activities amounted to PLN 221,830 thousand at the end of the analysed period (Table 4).

From the point of view of liquidity as the ability to pay current liabilities, the above tendencies are detrimental, but market practice justifies the statement that such financial structures are characteristic for companies with a strong competitive position *vis a vis* their suppliers, who extend mercantile credit to this strong buyer and thus finance its operations with the cheapest credit – deferred payment. At the same time, the strong competitive position ensures that even when the liabilities become overdue, suppliers do not charge penalty interest for fear of losing their selling market. Even though the mine did not have overdue liabilities in the analysed period, the degree to which it utilised mercantile credit from its suppliers was significant both in 2007 and 2008.

Table 2. Movements in net working capital and demand for it in 2007 [PLN'000]

	Jan. 2007	Feb. 2007	Mar. 2007	Apr. 2007	May 2007	Jun. 2007	Jul. 2007	Aug. 2007	Sep. 2007	Oct. 2007	Nov. 2007	Dec. 2007
Net working capital	-117,917.5	-115,373.5	-129,361.3	-141,131.2	-137,923.2	-261,697.7	-248,010.3	-231,781.5	-210,398.9	-299,026.5	-280,670.1	-296,496.7
Movement in NWC		*2,544.0*	*-13,987.8*	*-11,769.9*	*3,208.0*	*-123,774.5*	*13,687.4*	*16,228.8*	*21,382.6*	*-88,627.6*	*18,356.4*	*-15,826.6*
Demand for net working capital	-121,276.7	-121,963.9	-135,687.3	-147,522.0	-147,442.6	-270,066.4	-252,688.0	-232,053.1	-211,016.8	-299,199.4	-281,283.1	-301,078.5

Table 3. Funds flows for individual months of 2007 [PLN'000]

	Jan-Feb	*Feb-Mar*	*Mar-Apr*	*Apr-May*	*May-Jun*	*Jun-Jul*	*Jul-Aug*	*Aug-Sep*	*Sep-Oct*	*Oct-Nov*	*Nov-Dec*
Long-term capital change (+ growth, - fall)	-343.6	-6,741.3	800.8	7,743.1	-126,205.1	12,183.3	13,929.3	19,163.9	31,742.9	13,238.2	-14,924.2
Fixed assets change (+ growth, - fall)	2,887.6	-7,246.5	-12,570.7	-4,535.1	2,430.6	1,504.1	2,299.5	2,218.7	-120,370.5	5,118.2	-902.4
Change in NWC	*2,544.0*	*-13,987.8*	*-11,769.9*	*3,208.0*	*-123,774.5*	*13,687.4*	*16,228.8*	*21,382.6*	*-88,627.6*	*18,356.4*	*-15,826.6*
Change in inventories	631.1	7,981.8	6,904.9	12,042.4	5,914.7	6,849.8	2,780.0	-7,881.8	-3,915.6	-8,077.4	-7,964.4
Change in liabilities	3,706.7	-5,742.6	-1,814.1	-2,123.1	14,173.9	-11,126.8	-34.8	13,366.5	13,647.0	4,839.3	-5,100.5
Change in short-term investments	3,231.2	-264.4	64.8	3,128.6	-1,150.7	-3,691.0	-4,406.1	346.3	-445.0	440.1	3,968.8
Change in prepayments	4,167.7	-1,630.8	-1,415.4	-911.6	-915.0	-542.9	-1,088.7	-1,052.0	-717.6	989.3	455.3
Change in short-term provisions	-1,836.8	-2,746.0	-2,472.0	-2,353.7	-339.5	-980.2	-1,667.7	-1,121.2	-1,508.0	19,705.0	183.4
Change in short-term liabilities	-7,300.7	-11,595.1	-12,877.6	-6,584.2	-141,452.0	23,215.9	20,710.5	17,653.6	-95,674.2	665.7	-7,473.8
Change in accruals/prepayments	-55.2	9.3	-160.5	9.6	-5.9	-37.4	-64.4	71.2	-14.2	-205.6	104.6
Change in NWC	*2,544.0*	*-13,987.8*	*-11,769.9*	*3,208.0*	*-123,774.5*	*13,687.4*	*16,228.8*	*21,382.6*	*-88,627.6*	*18,356.4*	*-15,826.6*

Table 4. Movements in net working capital and demand for it in 2008 [PLN'000]

	Jan. 2008	Feb. 2008	Mar. 2008	Apr. 2008	May 2008	Jun. 2008	Jul. 2008	Aug. 2008	Sep. 2008
Net working capital	-265,701.9	-247,585.1	-228,183.9	-195,553.3	-175,634.0	-245,781.8	-259,725.9	-242,190.8	-221,830.8
Change in NWC	*30,794.8*	*18,116.8*	*19,401.2*	*32,630.6*	*19,919.3*	*-70,147.8*	*-13,944.1*	*17,535.1*	*20,360.0*
Demand for net working capital	-266,269.7	-248,106.9	-228,500.6	-195,745.3	-179,257.5	-246,550.8	-260,281.4	-242,676.9	-223,497.2

Table 5. Funds flows for individual months of 2008 [PLN'000]

	Dec-Jan	*Jan-Feb*	*Feb-Mar*	*Mar-Apr*	*Apr-May*	*May-Jun*	*Jun-Jul*	*Jul-Aug*	*Aug-Sep*
Long-term capital change (+ growth, - fall)	27,800.2	16,474.7	19,551.9	29,376.7	17,303.1	-73,154.6	-16,080.7	16,922.0	20,327.2
Fixed assets change (+ growth, - fall)	2,994.6	1,642.1	-150.7	3,253.9	2,616.2	3,006.8	2,136.6	613.1	32.8
Change in NWC	*30,794.8*	*18,116.8*	*19,401.2*	*32,630.6*	*19,919.3*	*-70,147.8*	*-13,944.1*	*17,535.1*	*20,360.0*
Change in inventories	-13,739.4	-1,908.7	411.4	-956.3	259.5	313.3	211.3	-1,134.6	884.0
Change in liabilities	13,236.9	-26,156.9	-1,071.6	3,477.7	-11,475.7	1,524.7	6,417.5	1,718.1	-263.3
Change in short-term investments	-4,014.0	-46.0	-205.1	-124.7	3,431.5	-2,854.5	-213.5	-69.4	1,180.3
Change in prepayments	-872.5	4,091.5	-1,502.4	392.7	-458.1	-592.9	-1,157.8	-381.7	-1,018.0
Change in short-term provisions	-2,602.0	-1,544.3	-6,955.1	-2,072.2	-2,093.2	-1,232.4	-5,210.0	-1,247.1	-1,758.2
Change in short-term liabilities	38,789.9	43,682.8	28,461.5	31,914.1	30,263.2	-67,320.0	-13,990.9	18,650.5	21,349.0
Change in accruals/prepayments	-4.1	-1.6	262.5	-0.7	-7.9	14.0	-0.7	-0.7	-13.8
Change in NWC	*30,794.8*	*18,116.8*	*19,401.2*	*32,630.6*	*19,919.3*	*-70,147.8*	*-13,944.1*	*17,535.1*	*20,360.0*

SUMMARY

In summary, it should be said that managing net working capital is a significant area in corporate financial management, also for companies operating in the mining industry. Financial efficiency should be due to the ability to pay current liabilities, while broadly understood financial balance must include primarily the balance between net working capital and the demand for it. In the dynamic approach, this translates into the ability to generate positive balances of cash from operations.

Research conducted at a mining company confirms the aggressive policy of managing capital and financing net working capital. Negative capital in the analysed period of 2007–2008 proves the strong competitive position of this company relative to its suppliers who finance the operations of the mine by giving it mercantile credit. Payment terms negotiated with suppliers allow the company to finance investment activities concerning fixed assets with current interest-free liabilities. In the analysed period, the mine did not use liabilities generating interest in the form of bank loans.

The identified tendencies are characteristic for companies operating within multi-facility structures having the nature of groups or holdings.

REFERENCES

[1] Bernstein L.: Analysis of Financial Statement. Homewood: Irwin 1993.

[2] Damodaran A.: Finanse korporacyjne. Wydawnictwo One Press, Gliwice 2007.

[3] Hawawini G., Viallet C.: Finanse menedżerskie. Polskie Wyd. Ekonomiczne, Warszawa 2007.
[4] Kusak A.: Płynność finansowa. Analiza i sterowanie. Wyd. Naukowe Wydziału Zarządzania Uniwersytetu Warszawskiego, 2006.
[5] Sierpińska M., Wędzki D.: Zarządzanie płynnością finansową przedsiębiorstwa. Wyd. Naukowe PWN, Warszawa 2002.
[6] Red. Pluta W.: Zarządzanie finansami firm – teoria i praktyka. Tom 2. Wyd. Akademii Ekonomicznej im. Oskara Langego, Wrocław 2003.
[7] Sierpińska M., Jachna T.: Ocena przedsiębiorstwa według standardów światowych. Wyd. Naukowe PWN, Warszawa 2004.
[8] Sierpińska M., Jachna T.: Metody podejmowania decyzji finansowych. Wyd. Naukowe PWN, Warszawa 2007.
[9] Wędzki D.: Strategie płynności finansowej przedsiębiorstwa. Oficyna Ekonomiczna, Kraków 2003.

International Mining Forum 2009, Sobczyk, Kicki & Saługa (eds) © 2009 Taylor & Francis Group, London, UK, ISBN 978-0-415-80428-8

Author Index